ACPL ITEM DISCARDED

3-16-76

Scanner-Monitor Servicing Guide

by

Robert G. Middleton

HOWARD W. SAMS & CO., INC.
INDIANAPOLIS · KANSAS CITY · NEW YORK

FIRST EDITION

FIRST PRINTING—1975

Copyright © 1975 by Howard W. Sams & Co., Inc., Indianapolis, Indiana 46268. Printed in the United States of America.

All rights reserved. Reproduction or use, without express permission, of editorial or pictorial content, in any manner, is prohibited. No patent liability is assumed with respect to the use of the information contained herein. While every precaution has been taken in the preparation of this book, the publisher assumes no responsibility for errors or omissions. Neither is any liability assumed for damages resulting from the use of the information contained herein.

International Standard Book Number: 0-672-21306-0
Library of Congress Catalog Card Number: 75-28963

Preface

Scanner-monitor receivers have become very popular, and the demand for technicians who understand this specialized branch of radio troubleshooting is increasing rapidly. Although much of the circuitry in a scanner monitor is the same as in conventional fm receivers, there are also some highly specialized networks associated with the scanning or automatic-tuning function. In turn, the scanner-monitor technician must become familiar with noise amplifiers and squelch gates, multivibrators, diode switching, counter and decoder/driver devices, and display devices such as light-emitting diodes and seven-segment readout indicators. This servicing guide starts at the beginning and proceeds step by step through the complete scanner-monitor system, with particular emphasis on specialized circuit action and troubleshooting.

The first chapter covers general considerations, to provide the reader with a suitable introduction to the general operation of a scanner-monitor receiver. In the second chapter, the various sections and subsections are discussed and their individual circuit actions are analyzed. Tuning circuits and alignment requirements are presented in the third chapter, with attention to appropriate test equipment. Next, single-conversion and double-conversion i-f sections are analyzed in the fourth chapter, again with attention to appropriate test equipment. Chapter 5 presents a comprehensive analysis of scan circuits and indicators, and notes the typical design variations that will be encountered. Chapter 6 covers troubleshooting procedures for the audio section. Specialized operating features of scanner-monitor receivers are discussed in Chapter 7. In all, seven detailed troubleshooting sections are included in this book.

Many illustrations are provided to clarify the technical points discussed in the text. Digital logic has been held to a minimum and is treated only to the extent required in practical servicing procedures. Typical variations in circuitry have been noted throughout the text in order to facilitate understanding and to make the troubleshooting procedures as general as possible. It is my firm belief and sincerest hope that this guidebook will be a valuable addition to the libraries of all present and future radio technicians, as well as a practical teaching tool in technical school programs.

ROBERT G. MIDDLETON

Contents

CHAPTER 1

GENERAL CONSIDERATIONS 7

General Arrangement of a Scanner Monitor—Scanner-Monitor Physical Layout—Basic Troubleshooting

CHAPTER 2

SCANNER-MONITOR SECTIONS AND SUBSECTIONS 17

Audio and Squelch Section—Scanning Section—Priority Oscillator Operation—Crystal-Oscillator Section—Audio Section—Power-Supply Section—Signal-Injection Tests—Signal-Tracing Tests—Troubleshooting Procedures

CHAPTER 3

TUNING CIRCUITS IN SCANNER-MONITOR RECEIVERS 31

Fixed-Tuned and Electronic-Tuned RF Circuitry—Oscillator Circuitry—RF Alignment Procedures—Troubleshooting Procedures

CHAPTER 4

I-F CIRCUITRY AND TROUBLE ANALYSIS 43

Typical I-F Circuitry—Semiconductor Testing—Troubleshooting Procedures

CHAPTER 5

SCAN CIRCUITS AND INDICATORS 55

Basic Scan-Circuit Tests—Channel Indicators—Priority Function—Voltage Specifications—Troubleshooting Procedures

CHAPTER 6

AUDIO SECTION TROUBLESHOOTING 69

    Basic Audio Circuit Tests—Precautions in Testing Audio Circuitry—DC Voltage Measurements—Troubleshooting Procedures

CHAPTER 7

SPECIALIZED SCANNER-MONITOR OPERATING FEATURES 81

    Transmitter Adjustments—Selective-Call Operation—Failure Modes of Digital ICs—Troubleshooting Procedures

INDEX 95

CHAPTER 1

General Considerations

Scanner-monitor (scanning monitor) radio receivers such as the one illustrated in Fig. 1-1 have automatic tuning capabilities in addition to manual tuning provisions. This is an example of an eight-channel, crystal-controlled, vhf scanning-monitor fm receiver that can be tuned to any selected 9-MHz segment of the 146- to 174-MHz frequency band. These are frequencies used by public-service organizations such as MARS (Military Affiliate Radio System), CAP (Civil Air Patrol), police departments, fire departments, ambulance services, and marine rescue services. Each of the eight channels is crystal-controlled.

Tuning can be accomplished either automatically or manually. When manual tuning is employed, the receiver will remain tuned to the selected channel. On the other hand, in the automatic mode of operation, the tuning system is automatically scanned through the programmed channels, and the receiver will lock in on the first channel with an active signal. The scanning monitor will then remain locked in on that particular channel as long as a carrier signal is provided. After the signal stops, the monitor delays scanning action for four seconds. Then, if there is still no signal output from the locked-in channel, the monitor will resume scanning. If the operator so desires, the monitor can be set to bypass one or more of the channels by pushing the appropriate frontpanel push button(s).

A priority channel (Channel 0), which has precedence over all other channels is provided in the example of Fig. 1-1. In other words, the monitor will automatically lock onto Channel 0 whenever a station is transmitting on that frequency, even though the monitor was previously locked onto some other channel. Digital readout is provided. The readout tube can be switched to display a channel number only when the monitor locks onto an incoming signal, or it can be switched to continuously display each channel number in sequence while the channels are being scanned. A jack is available on the rear panel for connection of an external speaker. Note that the frequency or station that is received on a particular channel depends on the crystal that is chosen for that channel. In some areas, it is illegal to use a crystal frequency that provides reception on police bands in a mobile receiver.

The vhf scanning monitor described above is an example of a comparatively elaborate receiver. Next, a simpler type of scanner monitor is illustrated in Fig. 1-2. This is an example of a four-channel pocket-scan fm monitor receiver. It has a frequency range from 144 to 175 MHz, and can be tuned to any selected 8-MHz segment within this

Courtesy Heath Co.

Fig. 1-1. An 8-channel, crystal-controlled, vhf scanning monitor.

Scanner-Monitor Servicing Guide

Courtesy Radio Shack

Fig. 1-2. A pocket-scan fm monitor receiver.

range. Note that the upper end of the 2-meter amateur band may be accommodated. Either automatic or manual tuning is available. Channel lockout switches are provided to permit "skipping" on one or more channels in the automatic mode. This is a useful feature, for example, when one of the crystal frequencies corresponds to National

Weather Service reception (162.55 or 162.40 MHz), because these broadcasts are continuous. Scanning speed is 10 channels per second. Light-emitting diode (LEDs) are utilized for the read-out in this example.

GENERAL ARRANGEMENT OF A SCANNER MONITOR

A block diagram for a typical vhf-fm scanner monitor is shown in Fig. 1-3. Double-conversion superheterodyne circuitry is utilized for maximum interference rejection, with i-f frequencies of 10.7 MHz and 455 kHz. Fixed broad-band tuning is employed in the rf and first mixer stages. A local oscillator operates from crystals, which are selected by the scanner (or selected manually), for optimum frequency stability and freedom from drift. An incoming fm signal has a deviation of approximately ±7 kHz; it heterodynes with the oscillator frequency to give an i-f signal with a 10.7-MHz center frequency. Next, this i-f signal is heterodyned with the output from a second crystal oscillator to produce the second i-f frequency of 455 kHz. (Simple scanner monitors omit the second i-f section.) A squelch circuit is included following the fm detector. This squelch circuit blocks the audio channel to suppress noise when there is no incoming signal. On the other hand, when a signal appears, the squelch circuit turns the audio channel on. The squelch-voltage level also starts or stops scanning action, as required.

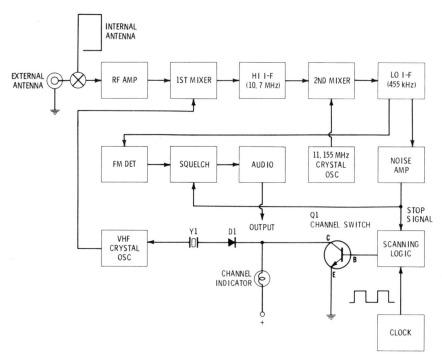

Fig. 1-3. Block diagram of a typical vhf-fm scanner monitor.

General Considerations

SYMBOL	FUNCTION
"AND" GATE (INPUTS → OUTPUT)	Output is "ON" only when all inputs are "ON". Output is "OFF" when any or all inputs are "OFF".
"NAND" GATE (INPUTS → OUTPUT)	Output is "OFF" only when all inputs are "ON". Output is "ON" when any or all inputs are "OFF".
"NOR" GATE (INPUTS → OUTPUT)	Output is "ON" only when all inputs are "OFF". Output is "OFF" when any or all inputs are "ON".
"OR" GATE (INPUTS → OUTPUT)	Output is "OFF" only when all inputs are "OFF". Output is "ON" when any or all of the inputs are "ON".

Fig. 1-4. Four basic gate symbols and their corresponding switching actions.

Next, the scanning-logic section in Fig. 1-3 comprises a group of electronic switches called *gates* which are opened or closed by the squelch voltage and also by the square-wave output from a multivibrator called the *clock*. In turn, the output voltage from the scanning-logic section is applied to a channel-switch transistor which operates as an electronic switch to turn the channel-indicator lamp on or off. Fig. 1-4 depicts four basic gate symbols and defines the switching action of each type of gate. The gates in this example have two inputs, but some gates may have as many as eight inputs. Regardless of the number of inputs, the principle of operation remains the same. For example, consider an AND gate that has three inputs. Its output will be "on" if all three inputs are "on" —its output will be "off" whenever one, two, or all three inputs are "off."

A block diagram for the scanning monitor illustrated in Fig. 1-1 is shown in Fig. 1-5. Receiver controls and functions are depicted in Fig. 1-6. Two rf stages are provided to obtain a high signal-to-noise ratio. The 10.7-MHz i-f output from the mixer passes through a crystal filter that consists of two monolithic crystal filters followed by a ceramic filter. Two integrated circuits in the i-f section provide amplification, limiting, and fm detection by a quadrature detector configuration. Practically all of the receiver selectivity is provided by the i-f circuitry. Audio output from the fm detector is applied to the squelch section which blocks passage of the audio signal into the audio amplifier when no signal is coming through the input circuits. In addition, the squelch section enables the digital-scanning section when there is no incoming signal.

A sudden increase in squelch voltage occurs when a signal arrives. In turn, the digital-scanning section is disabled and the audio amplifier is enabled. Thus, the receiver is locked onto the associated channel until the received signal stops. After the signal stops, there is a four second delay to avoid loss of communication in case the transmission happens to be briefly interrupted. Thereupon, scanning action resumes. In this example, when a signal has the receiver locked onto any channel other than the priority channel "0," a pulsing noise is audible in the background of the audio output. This pulsing is caused by the scanning of the priority channel for the presence of a signal. If the priority channel is locked out, the pulsing noise does not occur. This pulsing noise, when present, has a repetition rate of approximately four seconds. In case a priority signal appears, the monitor will break lock from any other channel and will lock onto channel "0" until the priority signal stops.

Note that the digital-scanning section in Fig. 1-5 operates by switching the eight crystal oscillators off and on in sequence at the same time that the display indicator is being driven. Each crystal is enabled or disabled by corresponding bias voltages

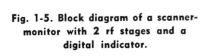

Fig. 1-5. Block diagram of a scanner-monitor with 2 rf stages and a digital indicator.

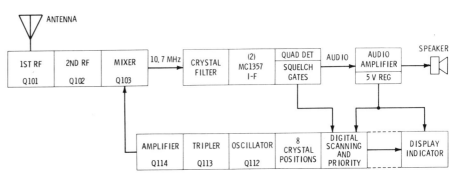

Courtesy Heath Co.

Scanner-Monitor Servicing Guide

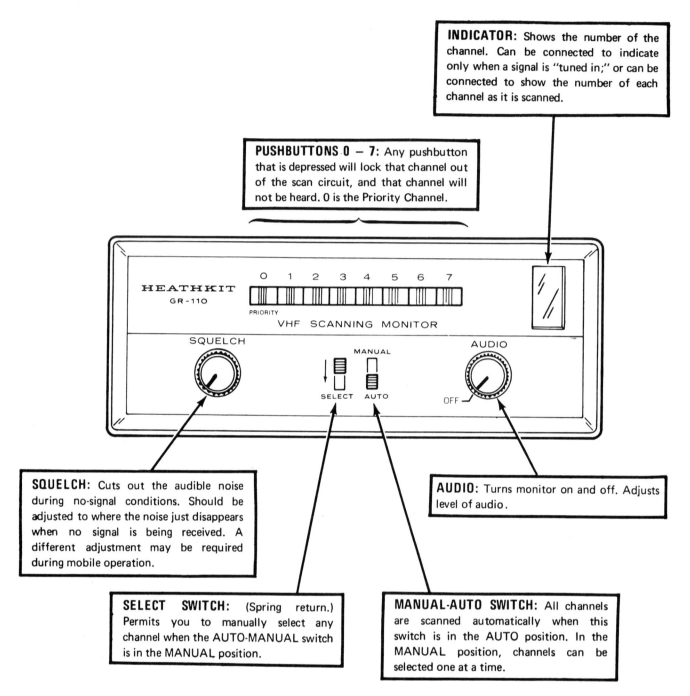

Fig. 1-6. Scanner-monitor controls and functions.

applied to associated diode switches. The oscillator output frequency is passed through a tripler stage, and the third-harmonic output is amplified and then fed to the mixer. Observe that a 7-segment display indicator is utilized in this example. One of the more recent and widely used types of display device is the 7-segment light-emitting diode (LED) depicted in Fig. 1-7. It operates from a 5-volt source and can be plugged into a standard IC socket. Display-device operation and servicing are explained in greater detail subsequently.

SCANNER-MONITOR PHYSICAL LAYOUT

A typical scanner-monitor layout is shown in Fig. 1-8. Two rf boards, one i-f board, a scan board, and a band-programming board are included. Crystal sockets are mounted on the rf

General Considerations

SYMBOL	OPERATION
(seven-segment display with segments a, b, c, d, e, f, g and DP)	All segments including the decimal point are common to pin 14. Pin 14 is normally connected to 5 volts DC. Each segment is then connected to ground through a 150 Ω, 1/4 watt resistor and Decoder/Driver. The Decoder/Driver causes the proper segments to light to display a number. CAUTION: Do not try to light the display without using the 150 Ω current limiting resistor, as the display will be ruined.

Courtesy GC Electronics

Fig. 1-7. Seven-segment light-emitting diode display device.

boards. In this example, the "200" board is a high-band vhf unit that covers the frequency range from 148 to 174 MHz. The "300" board is a low-band vhf unit that covers the frequency range from 30 to 50 MHz. A scanner monitor may employ a uhf board instead of one of the aforementioned vhf boards. A uhf board covers the frequency range from 450 to 470 MHz. Band-programming boards are connected as desired by the technician to provide scanning of a particular group of channels. As an illustration, Fig. 1-9 shows how plugs are connected in a 10-channel monitor to scan positions 1, 3, 8, and 10 on vhf (high-band vhf), and to scan positions 2, 4, 5, 6, 7, and 9 on lf (low-band vhf) or on uhf, depending upon which type of rf board has been installed.

Fig. 1-10 shows the locations of the oscillator crystals and the batteries in a typical pocket scanner monitor. Six battery cells are used, to provide

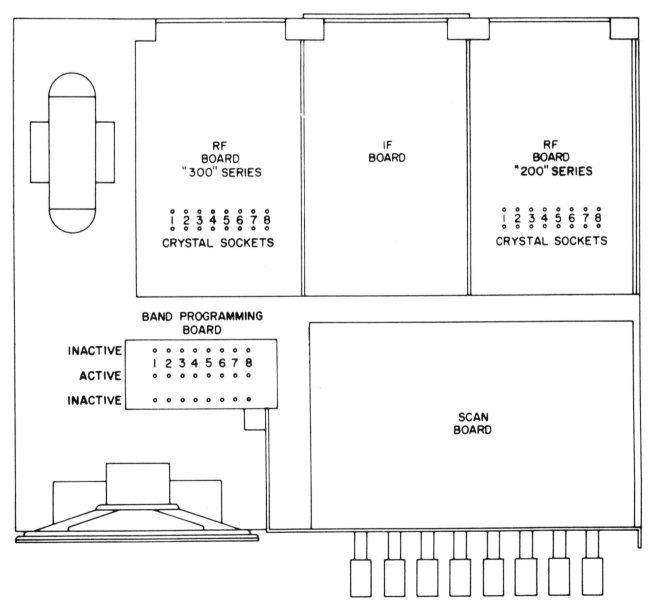

Fig. 1-8. Layout of high-band/low-band scanner monitor.

Scanner-Monitor Servicing Guide

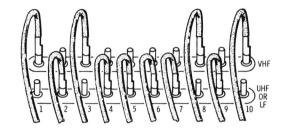

Fig. 1-9. Plug/connector arrangement on band-programming board for dual-band receiver.

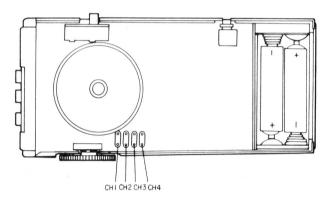

Fig. 1-10. Crystal and battery locations in a typical pocket scanner monitor.

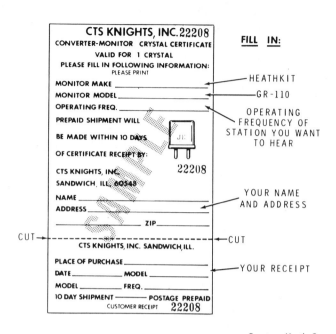

Fig. 1-11. A crystal certificate.

a source voltage of 9 volts. As a practical note, this type of monitor requires 1.5-volt cells—1.4-volt nickel-cadmium cells provide marginal source voltage and the monitor may not work on 8.4 volts.

Therefore, the batteries must be replaced with 1.5-volt cells, preferably of the rechargeable type. A battery jack is provided for recharging so that the batteries do not need to be removed from the monitor. Note that the crystal frequencies are very critical and that replacement crystals should be ordered in accordance with instructions in the scanner-monitor servicing data. It is poor practice

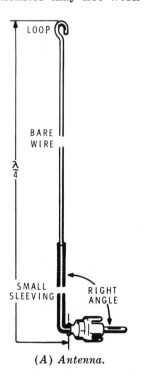

(A) Antenna.

FREQUENCY IN MHz	LENGTH $\frac{\lambda}{4}$	FREQUENCY IN MHz	LENGTH $\frac{\lambda}{4}$
146	20-1/4"	162	18-1/4"
147	20-1/8"	163	18-1/8"
148	19-7/8"	164	18"
149	19-3/4"	165	17-7/8"
150	19-5/8"	166	17-3/4"
151	19-1/2"	167	17-5/8"
152	19-3/8"	168	17-1/2"
153	19-1/4"	169	17-1/2"
154	19-1/8"	170	17-3/8"
155	19"	171	17-1/4"
156	18-7/8"	172	17-1/8"
157	18-3/4"	173	17"
158	18-5/8"	174	17"
159	18-1/2"		
160	18-1/2"		
161	18-3/8"		

NOTE: $\frac{\lambda}{4} = \frac{1}{4} \times$ WAVELENGTH (IN METERS)

ANTENNA CHART

(B) Antenna lengths for high-band vhf.

Courtesy Heath Co.

Fig. 1-12. Basic ¼-wavelength vertical antenna.

to attempt to replace a crystal with a substitute from another design of scanner monitor. A typical crystal certificate is shown in Fig. 1-11.

Most scanner monitors contain a telescoping monopole antenna or a built-in wire loop antenna, with provisions for plugging in an external antenna. Fig. 1-12 depicts a simple plug-in type of vertical antenna. For maximum sensitivity, the antenna wire should be cut to ¼ wavelength at the operating frequency. Thus, a 146-MHz antenna will be 20.25 inches in length, whereas a 174-MHz antenna will be 17 inches in length. Fig 1-13 shows

Fig. 1-13. Plug-in antennas for pocket scanners. Shorter antenna is for the uhf band.

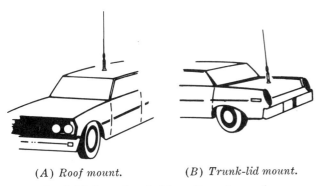

(A) Roof mount.   (B) Trunk-lid mount.

Fig. 1-14. Examples of vhf monitor antennas for mobile installations.

flexible plug-in antennas for pocket scanners. The vhf antennas shown in Fig. 1-14 are used for mobile installations. Such mobile antennas can be mounted on the top or the side of the trunk lid. RG-58/U coaxial cable is generally used to connect the antenna to the scanning monitor. A CB/vhf cowl-mount antenna may also be utilized for scanner-monitor receivers. This type of antenna has a maximum length of 50 inches, and is provided with a loading coil. Antennas are explained in greater detail subsequently.

A few scanner monitors are designed as transceivers. The same antenna is used for transmitting and for receiving. The arrangement for a typical scanner-monitor transceiver is shown in Fig. 1-15. Note that a transmit/receive (T/R) re-

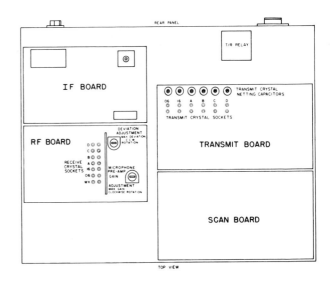

Courtesy Regency Electronics, Inc.

Fig. 1-15. Arrangement of a typical scanner-monitor transceiver.

lay is included for automatically switching the antenna to the receiver or to the transmitter. In this example, seven receiving crystal frequencies and six transmitting crystal frequencies are provided. The transmitter utilizes a narrow-band fm (nbfm) configuration. In some scanner-monitor communication systems, the transmitter is tone-coded. As depicted in Fig. 1-16, the tone code consists of either a low tone followed by a high tone, or a high tone followed by a low tone. In turn, the receiver contains a tone-decoder board, which causes the squelch circuit to open the audio channel when the appropriate tone sequence is received. On the other hand, a different tone sequence does not open the audio channel.

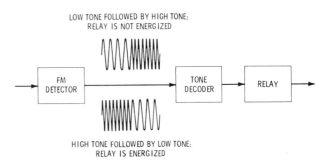

Fig. 1-16. Basic tone-decoder arrangement.

BASIC TROUBLESHOOTING

Some trouble symptoms are comparatively clear-cut and throw immediate suspicion on certain components or circuit sections. Other trouble

Scanner-Monitor Servicing Guide

symptoms are relatively indefinite and require various tests and measurements of the receiver circuitry to pinpoint the fault. The following examples of straightforward trouble symptoms and their causes provide a helpful introduction to scanner-monitor servicing.

1. No Sound Output and No Indicator Light

Probable causes for lack of sound output and no indicator light are as follows:

a. Dead battery, such as the 9-volt batteries shown in Fig. 1-17. (This is a block diagram for the scanner monitor illustrated in Fig. 1-2.)
b. Defective power switch. The power switch is mounted on the back of the volume control in the example of Fig. 1-17. Short-circuit the switch terminals for a quick check.
c. Short-circuit or open-circuit fault in the power circuit. Check with voltmeter and/or ohmmeter.
d. Battery connector has been replaced in wrong polarity; transistors and integrated circuits may have been damaged. Check polarity with a voltmeter and refer to the monitor servicing data.

2. No Sound Output, Indicator Lights All Right

Common causes for lack of sound output when the indicator lights operate normally are as follows:

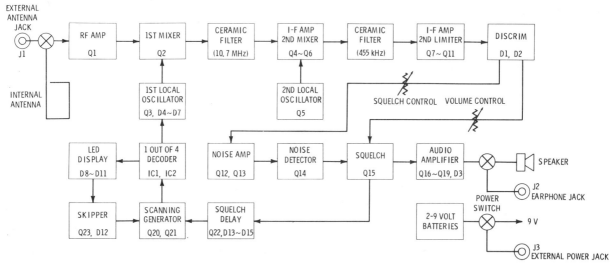

(A) Block diagram.

(B) Internal view.

(C) Foil side of circuit board.

Courtesy Radio Shack

Fig. 1-17. Pocket-scan fm monitor receiver.

a. Defective earphone jack (see Fig. 1-17). Make ohmmeter test.
b. Speaker defect; cross-check with earphone. Check continuity of voice coil.
c. Open volume control. Test with ohmmeter.
d. Defective squelch control (detailed subsequently).
e. Short circuit in antenna input section. (Open circuit can also cause the same symptom in some situations.)

3. No Sound Output With Earphone

Typical causes for no sound output with earphone, although the speaker operates normally, are as follows:

a. Defective earphone jack (Fig. 1-18). Make ohmmeter test.
b. Fault in earphone cord. Cross-check with another earphone.
c. Defective earphone. Cross-check with another earphone.
d. Earphone plug may be corroded or otherwise "insulated." Clean the plug thoroughly.

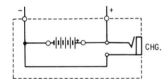

Fig. 1-18. Jack provided to recharge battery.

4. Noise Output, No Station-Signal Output

Probable causes of noise output, but no station-signal output, are as follows:

a. Weak battery. Check battery voltage under load with voltmeter.
b. Incorrect crystal plugged into oscillator section.
c. Defective crystal. Make a substitution test, if possible.
d. Faulty antenna connections (in some situations).

5. Weak Output

Common causes for weak output are as follows:

a. Weak battery. Check battery voltage under load with voltmeter.
b. Incorrect crystal plugged into channel socket.
c. Antenna defect. Make continuity check with ohmmeter.
d. Low field strength in the particular location (such as inside a building with metal construction).
e. Internal damage due to dropping the scanner monitor on a hard surface (this type of damage is detailed subsequently).

6. Distorted Output

Typical causes for distorted output are as follows:

a. Incorrect crystal or defective crystal in oscillator section. Check crystal requirements; make substitution test.
b. Severe reflections of the signal in a particular location. Try another location for comparison.
c. Speaker defect. Cross-check with earphone.
d. Transmitter malfunction; not likely, but possible. Cross-check with another scanner monitor.

7. Interference

Probable causes for interference in reception are as follows:

a. Excessively strong and closely spaced signals present in some urban and metropolitan areas. Use internal antenna instead of external antenna. Operate scanner monitor inside of a partially shielded location, such as an automobile.
b. Crystal not operating precisely on channel frequency. Replace with appropriate crystal.
c. Internal defect in i-f circuitry (this type of malfunction is explained in detail in a following chapter).

8. Short Battery Life

Probable causes of short battery life are as follows:

a. (Rechargeable battery.) Defective diode or other fault in charger. Check output voltage of charger under load. (See Fig. 1-17.)
b. (Rechargeable battery.) Battery may be nearing the end of its useful life. Try a new battery.
c. (Rechargeable battery.) Wrong type of battery may have been installed, such as 1.4-volt cells instead of 1.5-volt cells.
d. (Nonrechargeable battery.) Battery may have deteriorated on shelf. Try a battery from fresh stock.
e. (Nonrechargeable battery.) Leakage in

Scanner-Monitor Servicing Guide

power switch or power wiring. Check for battery current with milliammeter when power switch is open.

9. Noise in Mobile Operation

Reception of fm signals is usually not affected by automobile ignition noise. However, in some vehicles, noise suppression is required. A properly maintained and adjusted ignition system is helpful in noise reduction. Other means of noise reduction or elimination are as follows:

a. Install resistor spark plugs or TVRS (resistor-suppressor) ignition cable.
b. Install suppressor-resistor TVRS cable between distributor cap and ignition coil.
c. Install a 0.5-μF coaxial capacitor at the "A" terminal of the generator.
d. Alternators may require attention if the diodes become defective or the slip rings become dirty.
e. Install a 0.1-μF coaxial capacitor in the lead from the ignition switch to the coil. Keep capacitor close to coil terminal. Make sure that coil mounting bracket is well grounded.
f. Connect a 39-ohm resistor in series with a 0.01-μF ceramic capacitor between the regulator field terminal and ground.
g. Insert a 0.20-μF coaxial capacitor between the armature terminal and ground.
h. Install 0.5-μF capacitors with 200-volt rating from gauge terminals to ground.

CHAPTER 2

Scanner-Monitor Sections and Subsections

Scanner-monitor servicing generally requires an understanding of the various sections and subsections in the receiver. For example, it is essential to know the difference between the rf section, the mixer, the i-f section, the squelch section, the audio section, the scanning section, the oscillator section, the crystal-filter subsection, the frequency-multiplier subsection, the priority subsection, and the display section. Scanner-monitor servicing data are indispensable in this regard. Identification of sections and subsections starts with a block diagram, such as shown in Fig. 2-1. Components and devices such as integrated circuits, ceramic filters, and display indicators must be recognized, and their operating characteristics must be known. On the other hand, it is not necessary to be familiar with the internal construction of components such as an integrated circuit.

A schematic diagram for a scanning-monitor receiver is shown in Fig. 2-2. In this example, the rf section is mounted on the scan circuit board. Two broad-band rf amplifier stages are used. Dual-gate, diode-protected field-effect transistors are utilized in the rf and mixer stages. Fixed LC coupling is provided; these broad-band tuned circuits will respond to any antenna signal within a 9-MHz segment of the range from 146 to 174 MHz. The 9-MHz segment that has been chosen cannot be changed by the operator of the scanner monitor. If some other 9-MHz frequency segment is chosen, the broad-band rf circuits must be realigned by an experienced technician. Alignment procedures are explained in a following chapter.

Two input signals are applied to the mixer transistor, Q103. The amplified rf signal is applied to gate 1 (G1), and a cw voltage from the crystal oscillator is applied to G2. This crystal-oscillator frequency is 10.7-MHz lower than the center frequency of the incoming rf signal. Note that rf-amplifier transistors Q101 and Q102 operate in class

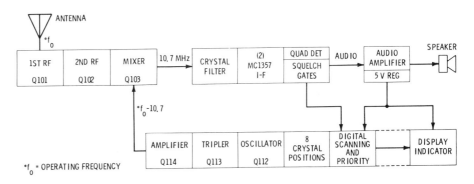

Fig. 2-1. Block diagram of a scanning-monitor receiver.

$*f_o$ = OPERATING FREQUENCY

Courtesy Heath Co.

Scanner-Monitor Servicing Guide

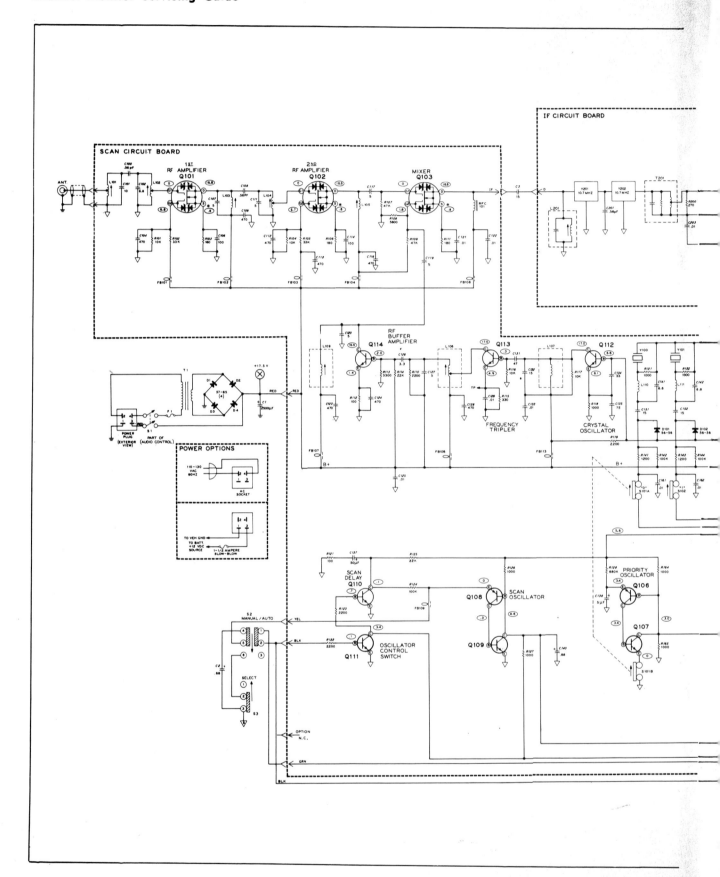

Fig. 2-2. Schematic diagram for

Scanner-Monitor Sections and Subsections

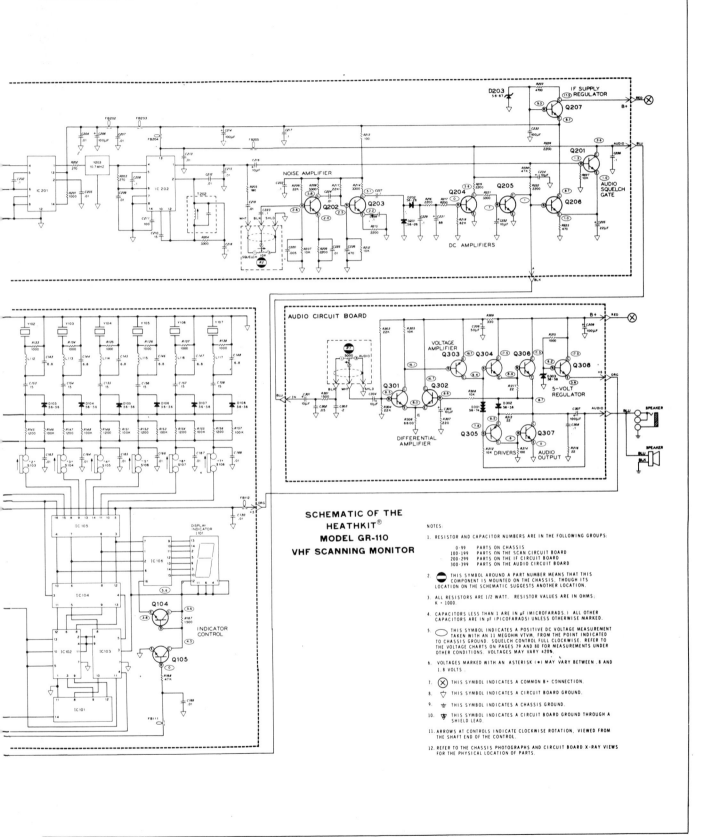

a scanning-monitor receiver.

Courtesy Heath Co.

A, with corresponding bias voltage applied to G2 of each transistor. On the other hand, mixer transistor Q103 operates essentially in class B, with considerably lower bias voltage on G2. In turn, a 10.7-MHz heterodyne output is obtained from Q13 and is fed to the input of the i-f section. Observe also that since the rf amplifier has broad-band response, a spectrum or "spread" of all the incoming signals is applied to the rf section. In other words, all of the selectivity for the scanner-monitor receiver must be provided by the i-f section. A few scanner monitors have varactor-tuned rf sections, as explained in Chapter 3. This design provides preselection prior to the i-f strip.

INTERMEDIATE-FREQUENCY NETWORK

With reference to Fig. 2-2, the spectrum of signals from the mixer section is passed through the monolithic crystal filters, Y201 and Y202, which have a 10.7-MHz center frequency. Tuned circuits L201 and T201 help to shape the i-f frequency-response curve. Most i-f bandpass filters in scanner monitors have monolithic construction. However, a few utilize discrete components. As an illustration, Fig. 2-3 shows a crystal-filter configuration with three crystals and associated capacitors. After amplification of the i-f signal through IC201, additional selectivity is provided by ce-

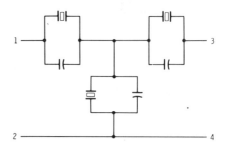

Fig. 2-3. A basic crystal-filter circuit with discrete components.

ramic filter Y203, which steepens the sides of the frequency-response curve. Normal bandwidth is approximately ±13.5 kHz at the −6-dB points (50% of maximum voltage amplitude points) on the response curve. Next, after further amplification followed by limiting and demodulation in IC-202, the signal is fed into the audio section. Limiting strips off any amplitude modulation that may accompany the fm signal. Quadrature detection is used in this example, with T202 functioning as the demodulator tank coil. Its adjustment determines the shape and symmetry of the S curve, which has frequency limits of ±15 kHz.

An understanding of i-f amplifier operation is facilitated by observing the internal circuitry of the integrated circuit (Fig. 2-4) used in the i-f section. The same IC is used in both the first and second i-f stages. As noted previously, the IC op-

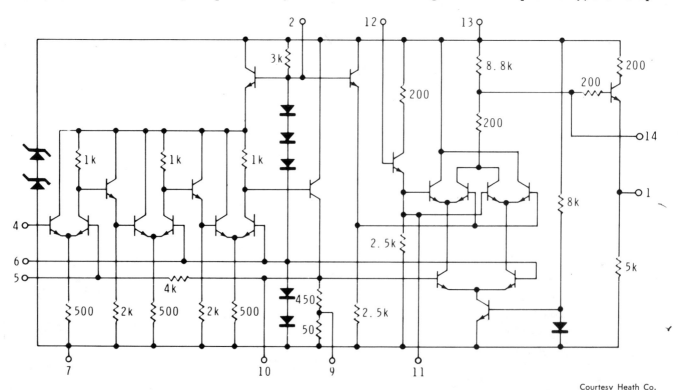

Courtesy Heath Co.

Fig. 2-4. Internal circuitry of integrated circuit used in the i-f amplifier.

Scanner-Monitor Sections and Subsections

erates as a class-A amplifier in the first stage, but is used as an amplifier limiter and quadrature detector in the second stage. From the viewpoint of troubleshooting, the important considerations are the input/output relationships of the IC and the normal dc terminal voltages. These topics are discussed in greater detail in following chapters.

AUDIO AND SQUELCH SECTIONS

The audio-frequency output at pin 1 of IC202 in Fig. 2-2 branches into two paths. One of these paths leads into the base of Q201, the preamplifier squelch gate transistor. The other path for the audio signal leads into the noise amplifier, which is a part of the squelch circuit. The purpose of the squelch circuit is to cut off the audio channel when no signal is coming through the i-f amplifier. Thereby, the receiver is automatically quieted (noise is suppressed) while the channels are being scanned. This is accomplished by passing the noise voltages through squelch control R2, amplifying the noise voltages through Q202 and Q203, rectifying the noise through the diode voltage-doubler circuit, filtering to pure dc, and then stepping up this control voltage through dc amplifiers Q204 and Q205. In turn, Q206 is a transistor gate which biases off (cuts off) Q201 when there is no incoming i-f signal.

Observe in Fig. 2-2 that there is no squelch action when the squelch control is turned fully clockwise. On the other hand, there is maximum squelch action when the squelch control is turned fully counterclockwise. In normal operation, the bias voltage at the emitter of Q201 is 1 volt when the squelch control is fully clockwise. Thus, Q201 is forward biased 0.3 volt, and amplifies in class-A operation. On the other hand, when the squelch control is turned fully counterclockwise, the bias voltage at the emitter of Q201 is 4.2 volts in normal operation. Consequently, Q201 is cut off, and no audio signal can pass into the audio section. Correct squelch action is obtained when the squelch control is adjusted so that the audio channel is just barely quieted while the scanner is operating. Then, when an i-f signal appears, the noise from the i-f amplifier ceases because the limiter becomes operative. In turn, the bias voltage on the emitter of Q201 falls to 1 volt, and the audio signal is amplified by Q201 and fed into the audio circuit board.

Keep in mind that some squelch controls are connected so that the audio channel is cut off when the control is turned fully clockwise. A technician unfamiliar with a particular receiver could be misled into believing that a receiver defect is present because he assumed that the squelch control operation is normally opposite. It might seem to the beginner that a voice waveform would cut off the audio channel in the same way that a noise waveform does. However, there is a basic difference between these waveforms, as depicted in Fig. 2-5. Most of the power in a voice waveform is in the lower audio frequencies. Therefore, the average voltage of a voice waveform is comparatively small at the higher frequencies. On the other hand, the average voltage of a noise waveform is the same at all frequencies within the passband of the receiver. Note in Fig. 2-5 that a 10-μF coupling capacitor is used at the input of the audio amplifier, whereas a 0.01-μF coupling capacitor is used at the input of the noise amplifier. Therefore, the noise amplifier responds chiefly to high frequencies, and a voice waveform normally produces very little output from the noise amplifier. However, a noise waveform produces a substantial output from the noise amplifier.

Observe in Fig. 2-2 that the dc control voltage from Q205 branches into the scanning section, as well as into Q206. Thus, when an i-f signal appears and the audio amplifier is enabled, the control voltage stops the scan oscillator at the same time. Thereby, the receiver is held on the active channel as the audio signal passes into the audio circuit board. Scanning resumes when the i-f signal ceases.

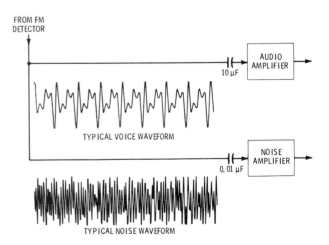

Fig. 2-5. Noise amplifier has comparatively great high-frequency response, and little low-frequency response.

SCANNING SECTION

With reference to Fig. 2-2, the squelch gate signal from Q205 is coupled to the base of transistor

Scanner-Monitor Servicing Guide

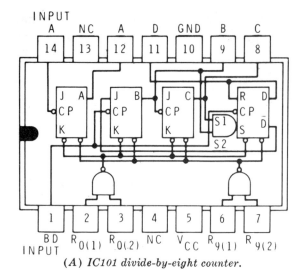

(A) IC101 divide-by-eight counter.

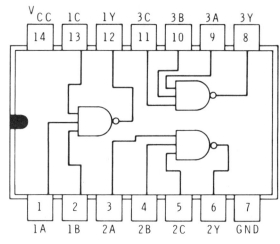

(B) IC102 and IC103 gating circuits.

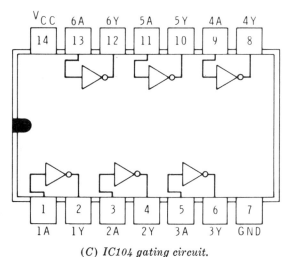

(C) IC104 gating circuit.

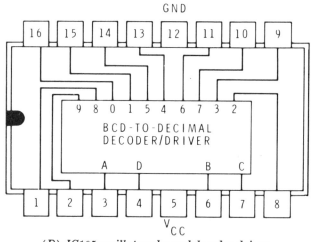

(D) IC105 oscillator channel decoder-driver.

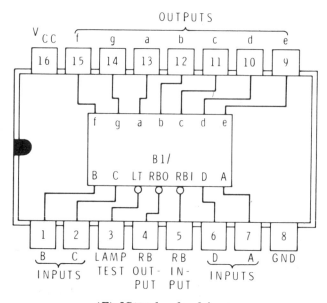

(E) IC106 decoder-driver.

Courtesy Heath Co.

Fig. 2-6. Basic data for the scan-section integrated circuits.

Q111. This transistor functions as a control switch, to start or stop the action of scan oscillator transistors Q108 and Q109. A short response delay is produced by RC circuitry in the Q110 scan-delay subsection. Thus, when a received signal ceases, a delay of approximately 4 seconds takes place before the scan oscillator transistors, Q108 and Q109, start the scanning action again. Fig. 2-6 shows basic data for the integrated circuits used in this example. Integrated circuit IC101 is called a divide-by-eight counter. It consists of a chain of flip-flops (bistable multivibrators). A chain of flip-flops counts pulses and divides as depicted in Fig. 2-7. The flip-flops trigger one another in succession. Counting takes place in terms of binary numbers. For example, binary and decimal numbers are related as follows:

Decimal Number	Binary Number
0	0000
1	0001
2	0010
3	0011
4	0100
5	0101
6	0110
7	0111
8	1000
9	1001
10	1010

Fig. 2-7 shows how the output from flip-flop 1 indicates the binary number 0 or 1, how the output from flip-flop 2 indicates the binary number 0 or 2, and how the output from flip-flop 3 indicates the binary number 0 or 4. A flip-flop can be compared with a spdt push-button switch connected to two lamps. When the button is pushed and released, one of the lamps will turn "on" and the other lamp will turn "off." If the button is again pushed and released, the lamp that was "off" will now turn "on," and the lamp that was "on" will turn "off." There are many different types of flip-flops, but they all operate basically the same. The difference between them is in the number and types of inputs. Fig. 2-8 shows how a flip-flop has "on" and "off" outputs called Q and $\overline{Q}$ (NOT Q).

A latch is a flip-flop that is used to store Q or $\overline{Q}$ information until it is needed. The latch has an extra input which is often called the latch input. Its symbol and functions are shown in Fig. 2-9. When the latch is turned on, the high or low voltage at the input will be transferred to the Q output. On the other hand, when the latch is turned

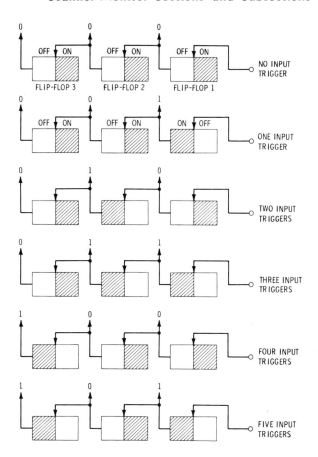

Fig. 2-7. Development of binary numbers by a flip-flop chain.

SYMBOL	FUNCTION
INPUT — FLIP-FLOP — Q OUTPUTS — $\overline{Q}$	"Q" output is "ON" when "$\overline{Q}$" output is "OFF". "Q" output is "OFF" when "$\overline{Q}$" output is "ON". Outputs change when the input is turned "OFF".

Fig. 2-8. Basic flip-flop symbol and function.

SYMBOL	FUNCTIONS
INPUTS — LATCH — Q OUTPUTS — $\overline{Q}$	"Q" output will follow the information input. "$\overline{Q}$" output will always be the opposite of the "Q" output.

Fig. 2-9. Latch-type flip-flop and functions.

off, the high or low voltage that was at the input at the time that the latch was turned off, will be stored on the Q output. The flip-flops in IC201 (Fig. 2-6A) are called JK flip-flops. A JK flip-flop has one more input than a latch, and its output will change or remain unchanged depending upon the high or low voltage states of the J and K inputs.

Scanner-Monitor Servicing Guide

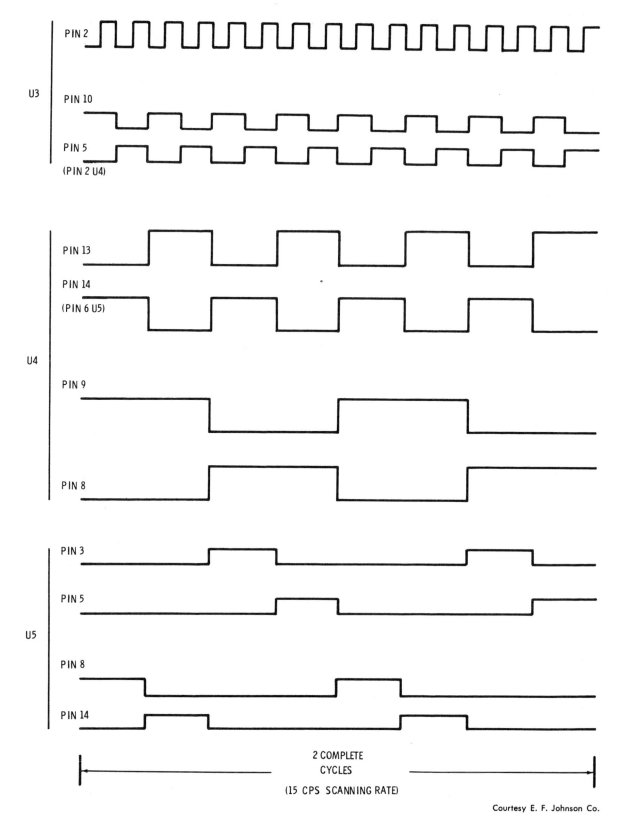

Fig. 2-10. Typical waveforms at active terminals of a digital-scanning IC.

Courtesy E. F. Johnson Co.

Integrated circuit IC201 also contains two NAND gates and an AND gate, as discussed in the previous chapter.

Oscilloscope tests at the active terminals of digital-scanning ICs normally display waveforms such as those shown in Fig. 2-10. The peak-to-peak amplitude of the waveforms is approximately 4 volts p-p in normal operation. Next, with reference to Fig. 2-6B, IC102 and IC10B contain NAND gates, as explained in the preceding chapter. Integrated circuit IC104 contains inverters. As shown in Fig. 2-11, an inverter is an amplifier that produces an inverted output. Integrated circuit IC105 contains a decoder-driver network for switching the oscillator channels off and on. A binary-coded decimal (bcd) decoder configuration is utilized. Integrated circuit IC106 is a decoder-driver network for operating the seven-segment display indicator. The preceding functions are summarized in Fig. 2-12, and will be considered in more detail in following chapters.

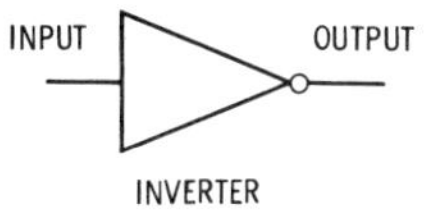

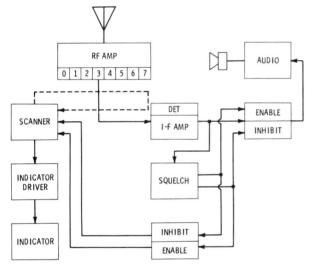

Fig. 2-11. Inverter symbol and functions.

Fig. 2-12. Summary of basic monitor scanning action.

PRIORITY OSCILLATOR OPERATION

With reference to Fig. 2-2, when a signal has the monitor locked on any channel other than priority channel "0," the priority oscillator will cause a pulsing noise every four seconds in the audio output, due to scanning of the priority channel for the presence of a signal. When integrated circuit IC101 stops counting and locks onto a channel, the priority oscillator, consisting of transistors Q106 and Q107, will pulse an input of each gate in IC102 and one gate in IC103 to ground. Pulsing causes the outputs of three input NAND gates to go "low." These output signals are then inverted by IC104 to a "high" (logic 1) and coupled to the inputs of IC105 and IC106. When all of their inputs are "high" (logic 1), a display indication of "0" will be produced by decoder-drivers IC105 and IC106.

This sequence occurs approximately every four seconds and has a duration of 20 milliseconds. During this 20 milliseconds, the "0" channel crystal oscillates. If an incoming signal appears, the scanning circuits lock onto the signal by using the squelch voltage to trigger pin 4 of IC103, which is an input of a 3-input NAND gate. In turn, a voltage is coupled from the "0" output through an inverter to produce a "high" on pin 5 of IC103. This ensures that the monitor will stay locked onto the priority channel when a priority signal is present.

CRYSTAL-OSCILLATOR SECTION

With reference to Fig. 2-2, the crystal oscillator comprising Q112 and associated circuitry operates in a Colpitts configuration. A quarter-wave impedance-transforming network is connected in series with the enabled crystal and the base of Q112. Note that the diodes in the crystal branches operate as electronic switches to provide an rf path between the crystal and the base of Q112; all diodes except one are normally biased off. Bias conditions are determined by the output levels from IC105. Observe that Q113 operates as a frequency tripler, and the third harmonic of the crystal is amplified by Q114. In turn, the output from Q114 is fed to the mixer transistor, Q103. The frequency of the Q114 output is usually 10.7 MHz below the rf signal frequency. However, when interference is a problem, the technician may employ a crystal frequency that provides mixer operation 10.7 MHz above the rf signal frequency.

AUDIO SECTION

A differential-input stage is utilized in the audio section (Fig. 2-2) for combining the input signal in Q301 with the negative-feedback signal in Q302. Negative feedback normally reduces audio distortion, as shown in Fig. 2-13. The signal at the collector of Q301 is coupled to the base of the constant-current voltage-amplifier transistor, Q303. In turn, the amplified signal is coupled to the base of Q304 and through diode D301 to the base of transistor Q305. The constant-voltage drop across diode D301 provides a bias voltage for the driver and output stages. This voltage is applied across

Scanner-Monitor Servicing Guide

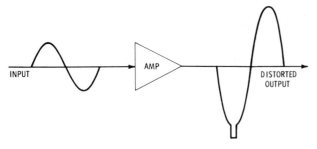

(A) Amplifier operation without feedback.

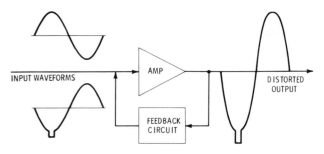

(B) Input waveforms with negative feedback.

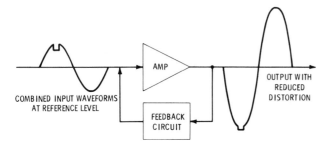

(C) Reduced output distortion resulting from combined input waveforms.

Fig. 2-13. Negative-feedback process in the audio amplifier.

the Q304 and Q305 base-to-emitter junctions that are connected in the circuit with diode D302. Resistor R311 drops the voltage sufficiently to keep both of these transistors (Q304 and Q305) biased "on" at all times in normal operation. Diode D302 limits the amount of voltage that will develop across resistor R311, thereby limiting the amount of current drawn by the output transistors (Q306 and Q307).

Driver transistors Q304 and Q305 and output transistors Q306 and Q307 form a quasi-complementary output section. In other words, the output transistors operate in class AB. A positive-going signal applied to the base of Q304 causes the transistor to conduct. This, in turn, causes output transistor Q306 to conduct. Negative-going signals cause transistors Q305 and Q307 to conduct. Thus, the current through the speaker voice coil and coupling capacitor C307 alternately goes positive and negative. To prevent motorboating and oscillation of the audio section in this example, the phase of the output signal must be shifted. This is the function of R316 and C308. Thus, in case the amplifier becomes unstable, C308 should be checked for an "open."

POWER-SUPPLY SECTION

With reference to Fig. 2-2, the power supply in this example consists of a dual-input power supply, with provision for a 13.8-volt dc source or a 120-volt ac source. When a 13.8-volt dc source is used, the power is routed through on-off switch S1 to filter capacitor C1, and thence to all of the circuits which operate on a supply voltage range of 13.8 to 17.5 volts. When a 120-volt ac power source is utilized, power is routed through on-off switch S1 and through fuse F1 to the primary of power transformer T1, which steps the voltage down to 17.5 volts. In turn, bridge-rectifier diodes D1 through D4 rectify the secondary voltage which is then filtered by C1. Note that the 5-volt power source for the scan circuits is regulated by Q308 and its associated circuitry. Also, the 13.8-volt supply is regulated by transistor Q207 and its associated circuitry.

SIGNAL-INJECTION TESTS

Signal injection is a very useful preliminary troubleshooting procedure. Although specialized signal generators can be utilized, a simple noise-injector probe such as that illustrated in Fig. 2-14 will serve for quick checks. This is a transistorized pulse generator that produces sharp pulses at an audio-frequency rate. These pulses have a wide range of harmonics. In turn, the probe can be used for signal-injection tests of the audio section, i-f section, and rf section of a receiver. Of course, the receiver must be unsquelched when the test signal is being injected into a circuit prior to the squelch-gate transistor. After the approximate trouble area has been localized by signal-injection tests, dc voltage measurements often suffice to pinpoint

Scanner-Monitor Sections and Subsections

SIGNAL TRACING TESTS

Technicians sometimes prefer to make signal-tracing tests instead of signal-substitution tests in preliminary troubleshooting procedures. A signal tracer such as the one illustrated in Fig. 2-15 is useful for signal-tracing the i-f and audio sections of a scanner monitor. Either a station signal or a signal generator can be used to provide a test indication. Note that a conventional signal tracer responds only to an rf signal with amplitude modulation or to an audio signal. Therefore, if a station signal is used, the signal tracer will respond only to minor irregularities and noise pulses in the i-f section. However, the signal tracer will respond to the audio information in a station signal following fm detection. Note also that although an a-m signal generator provides a suitable indication for i-f tests, the amplitude modulation will be eliminated by the limiter. Therefore, the audio section does not produce any output unless there is appreciable incidental fm in the test signal. In practice, most

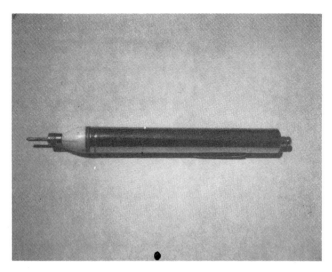

Fig. 2-14. Noise-injector probe is useful for preliminiary troubleshooting.

defective components. In case of doubt, resistance measurements will be helpful.

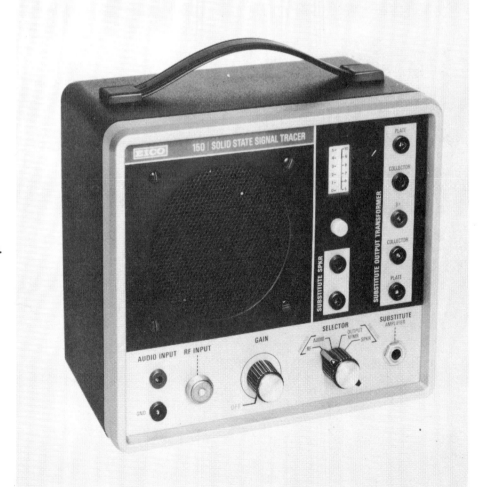

Fig. 2-15. A typical signal tracer.

Courtesy EICO Electronic Instrument Co., Inc.

service-type a-m generators have substantial incidental frequency modulation when operated at a comparatively high percentage of modulation.

Scanner waveform checks may be included in the general classification of signal-tracing tests. Most technicians do not make detailed observations and measurements of scanner waveforms, although it is often quite helpful to determine whether a particular scanning waveform is present or absent. Even if the waveform is present, it may fail to actuate the intended device if the waveform amplitude is subnormal. Therefore, it is good practice to use a calibrated scope to check the digital waveform amplitudes. Technicians who wish to make comprehensive tests in digital circuitry may use a dual-trace triggered-sweep oscilloscope, such as that shown in Fig. 2-16. This type of scope can display input and output waveforms simultaneously, showing their timing relationships. Similarly, the time relationships of any pair of digital waveforms in the scanning section can be displayed.

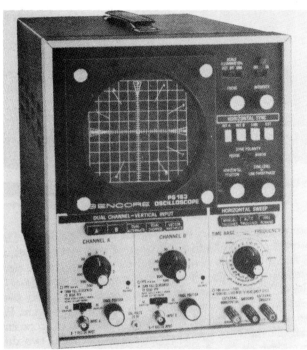

Courtesy Sencore

Fig. 2-16. A dual-trace triggered-sweep oscilloscope.

TROUBLESHOOTING PROCEDURES

1. Completely Dead Receiver

Probable causes for a completely dead scanner monitor are as follows:

a. Blown fuse in ac operation, such as F1 in Fig. 2-2. A new fuse is likely to blow in a short time; check the power supply for a defective component, such as a leaky filter capacitor or a shorted rectifier diode.

b. Blown fuse when operating from 12-volt dc supply, as shown in Fig. 2-2. If a new fuse blows, check the filter capacitor. In case the filter capacitor is all right and the current drain of the scanner monitor is excessive; look for a short-circuited power transistor.

c. Power plug, cord, or socket may be defective. Check with ohmmeter. Power switch occasionally becomes erratic or inoperative.

d. Scanner monitor may have been extensively damaged by connection to a dc source in wrong polarity. In such a case, the ICs and transistors must be systematically checked.

e. Scanner monitor may have been dropped and mechanically damaged. Look for cracked circuit boards.

2. No Output From I-F Section

If a signal-injection test with a noise generator indicates that there is no output from the i-f section but that the audio section is operative, the most likely trouble causes are as follows:

a. In case there is output from the second stage, check the first-stage IC, such as IC201 in Fig. 2-2. In this example, IC201 and IC202 can be interchanged for a substitution test.

b. If the ICs are all right, check the associated capacitors, such as C201 through C205 in Fig. 2-2.

c. Measure the i-f supply voltage; a cold-soldered joint, for example, could block supply voltage to an i-f stage.

d. Inspect the IC socket contacts carefully to make certain that all pins make good connection to the terminals.

e. Check to see whether an IC may have been reversed (turned around).

3. No Audio Output (Scanning Action Normal)

When there is no audio output in response to signal-injection tests, and scanning action is normal, probable causes are as follows:

a. Spare phono plug left inserted into external speaker jack.

b. Low or no supply voltage to audio circuit board; measure voltage.

c. Defective transistor in audio section; check terminal voltages of the transistors. A voltage-shifting diode might also become defective.

d. Leaky or open capacitor in audio section (electrolytic capacitors are common offenders).
e. Defective speaker—not likely, but possible; cross-check with test speaker.

4. Low Sensitivity

Low sensitivity can be caused by a malfunction in the rf, i-f, or audio sections. Therefore, it is helpful to make signal-injection comparison tests on a similar scanner monitor (if possible). Typical causes of low sensitivity are as follows:

a. In case of weak output from injected rf signals, measure the rf supply voltage.
b. Check antenna connections and circuit continuity.
c. Measure transistor terminal voltages; a transistor may be defective, or an associated capacitor might be leaky.
d. If crystals have been changed, rf circuits could need realignment, as detailed subsequently.
e. In case of weak output from injected i-f signals, measure the i-f supply voltage.
f. Check IC terminal voltages, if specified; otherwise, make substitution test. In case of discrete devices, check transistor terminal voltages.
g. In case of weak output from injected audio signals, measure the audio section supply voltage.
h. Measure transistor terminal voltages.
i. Check audio-circuit capacitors, such as C307 in Fig. 2-2.

5. Monitor Does Not Scan Automatically

When a monitor does not scan automatically, the squelch control may be incorrectly set. Otherwise, the following possibilities should be checked out:

a. A defective component in the noise amplifier or in the noise detector circuit. The squelch does not operate in this case.
b. Signal does not get through because of malfunction in the rf or i-f amplifier (see preceding topics).
c. If the squelch operates normally but the monitor does not scan, measure voltages at terminals of logic ICs and transistors.
d. Check capacitors in scanning section; also check the manual/auto switch.
e. Supply-voltage regulator may be defective—Q308 and associated components in Fig. 2-2, for example.

6. Manual Selector Does Not Operate

Typical causes for an inoperative manual selector are as follows:

a. Switch defective due to excessive wear.
b. Oscillator control transistor faulty; measure terminal voltages.
c. Deteriorated integrated circuit, such as IC-103 in Fig. 2-2.
d. Defective capacitor in manual selector circuitry.
e. Off-value resistor in manual selector circuit—not likely, but possible.

7. Audio Distortion

When audio distortion occurs, the trouble is not necessarily in the audio section. If a tape player or other audio source is available, the output can be injected into the audio section of the scanner monitor to determine whether the trouble is in the audio section or elsewhere. Common causes for audio distortion are as follows:

a. Defective crystal in oscillator section.
b. Crystal filter misaligned in i-f section, as explained in greater detail in a following chapter.
c. Defective quadrature coil, such as T202 in Fig. 2-2.
d. Incorrect transistor bias voltage in audio-amplifier circuitry.
e. Faulty transistor in push-pull audio output circuit.

8. Unstable Operation

Probable causes of unstable operation are:

a. Open stabilizing capacitor, such as C308 in Fig. 2-2.
b. Open decoupling capacitor, such as C204, C206, C207, C213, C214, C217, or C236 in Fig. 2-2.
c. Defective transistor or diode in regulator circuit. Monitor the regulated voltages.
d. Worn and noisy volume control or squelch control.
e. Marginal crystal—not likely, but possible.
f. If scanner-monitor receiver has been subjected to excessive humidity, dry it out thoroughly before testing and troubleshooting.

CHAPTER 3

Tuning Circuits in Scanner-Monitor Receivers

The rf tuning circuits in scanner-monitor receivers may cover one, two, or three bands. As an illustration, the receiver illustrated in Fig. 3-1 accommodates 16 channels selected from the 30–50 MHz vhf band, the 148-174 MHz vhf band, and the 450–470 MHz uhf band. Each channel is 27 kHz in width. Most scanner-monitor receivers have fixed-tuned rf circuits. For example, Fig. 3-2 shows a typical frequency coverage for the tuned rf circuits in a uhf scanner monitor. The uhf range extends from 450 to 470 MHz, and the frequency coverage for the tuned rf circuits in this example is from 455 to 465 MHz. In turn, this frequency coverage accommodates 370 channels with reasonable sensitivity. Note that the tuned rf circuits will respond to frequencies outside the limits from A to B, but the output will be comparatively weak.

Maximum rf sensitivity in Fig. 3-2, of course, is obtained at a frequency of 460 MHz, which corresponds to the peak of the frequency-response curve. Note that the rf circuits are adjustable. In other words, if it were desired to obtain maximum sensitivity at 450 MHz, the rf circuits could be realigned to provide peak output at this frequency. Again, if maximum sensitivity were desired at 470 MHz, the rf circuits could be adjusted to provide peak output at this frequency. Therefore, it is general practice to align the tuned rf circuits in accordance with the channel frequencies of interest to the operator in a particular locality. Alignment procedure is explained in detail in the latter portion of this chapter.

The frequency coverage for tuned rf circuits in the uhf range is greater than for those in the high vhf range, as seen by comparing Fig. 3-2 with Fig. 3-3. The high vhf range extends from 148 MHz to 174 MHz, with a typical rf tuned-circuit coverage of 8 MHz. In other words, the useful rf sensitivity occupies 31% of the high vhf range. On the other hand, as seen in Fig. 3-2, the useful

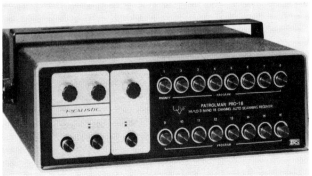

Courtesy Radio Shack

Fig. 3-1. A scanner-monitor receiver that covers three bands.

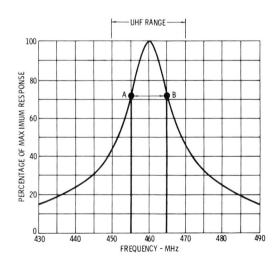

Fig. 3-2. Frequency coverage of uhf tuned circuit is 10 MHz for maximum sensitivity.

Scanner-Monitor Servicing Guide

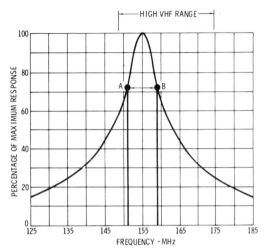

Fig. 3-3. Frequency range of high vhf tuned rf circuit is 8 MHz for maximum coverage.

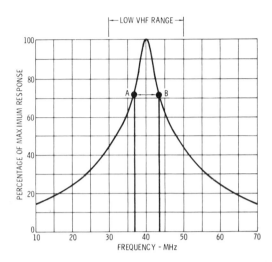

Fig. 3-4. Frequency range of low vhf tuned rf circuit is 6 MHz for maximum coverage.

rf sensitivity occupies 50% of the 450–470 MHz uhf range. Next, with reference to Fig. 3-4, the low vhf range extends from 30 to 50 MHz, with a typical tuned-circuit coverage of 6 MHz. In other words, the useful rf sensitivity occupies 30% of the low vhf range. From a practical viewpoint, this means that realignment of the tuned rf circuits is more likely to be desirable when oscillator crystals (channels) in the low or high vhf ranges are being changed, than when oscillator crystals in the uhf range are being changed.

FIXED-TUNED AND ELECTRONIC-TUNED RF CIRCUITRY

Although most scanner monitors have one stage of rf amplification, a few receivers use two stages. Fig. 3-5 shows an example of a single-stage, fixed-tuned rf amplifier for the 25–50 MHz (low vhf) range. It uses two rf transformers with adjustable cores. In turn, the rf amplifier can be aligned for optimum sensitivity over the desired band of frequencies for a particular locality. Observe also in this example that capacitors C102 through C106A employ different values in the 25–32.5 MHz, 32.5–41 MHz, and 4–50 MHz frequency intervals. With reference to Fig. 3-4, this example of an alignment condition is for a band of frequencies from 37 MHz to 43 MHz. Peak response occurs at 40 MHz. Note that a 50-MHz channel signal will be attenuated to 45%, or to less than half of its voltage at the antenna terminal. In other words, unless a 50-MHz signal were comparatively strong, it might not be received satisfactorily with this alignment condition. Similarly, a 30-MHz channel signal will be attenuated to 45%. On the other hand, no attenuation occurs at 40 MHz.

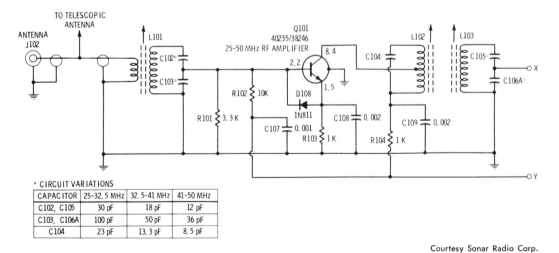

Courtesy Sonar Radio Corp.

Fig. 3-5. Example of a single-stage, fixed-tuned rf circuit.

Tuning Circuits in Scanner-Monitor Receivers

The tuned rf circuits in Fig. 3-5 could be realigned for peak response at 30 MHz, or at 50 MHz, if desired. In case a 30-MHz channel frequency were of greatest importance in a particular locality, it would be desirable to align the rf circuits for peak response at 30 MHz. In turn, a 50-MHz channel signal would be attenuated to 25% of its amplitude at the antenna terminal. Further, a 40-MHz channel signal would be attenuated to 45% of its incoming amplitude. Note that alignment adjustments are usually stable and do not change or drift excessively unless a component defect occurs. Therefore, the general rule is to check alignment last, after all other troubleshooting procedures are completed. Transistor terminal voltages are specified in Fig. 3-5. Preliminary troubleshooting usually consists of measuring these voltages. Although an incorrect voltage is often caused by a defective transistor or diode, leaky or shorted capacitors can also be responsible.

Greater sensitivity is provided at frequencies considerably removed from the peak rf response frequency by the use of two rf stages, as exemplified in Fig. 3-6. The sensitivity at peak rf response depends upon the alignment bandwidth. In other words, there are five tuned rf circuits provided in Fig. 3-6, and the peak rf response will be quite high when all the circuits are resonated at the same frequency. However, with a small amount of stagger tuning, the peak rf response may be no greater than when one rf stage is used, but the 3-dB bandwidth will then be considerably greater than when only one rf stage is used. That is, greater sensitivity is thereby provided at frequencies considerably removed from the peak rf response frequency. Transistor Q103 operates as a mixer stage in Fig. 3-6, and Q104 operates as the local oscillator, with emitter injection to the mixer. T301 and T302 are mixer-output tuned circuits operating at 10.7 MHz.

Peak response on all channels is provided by more elaborate scanner-monitor receivers that include an automatic rf tuning system (electronic tuning), as exemplified in Fig. 3-7. This arrangement uses an integrated circuit and associated components and devices to track the rf circuitry with the oscillator frequency as the scanning process takes place. The tuned rf circuits are resonated to the correct frequency at each step of the scan-

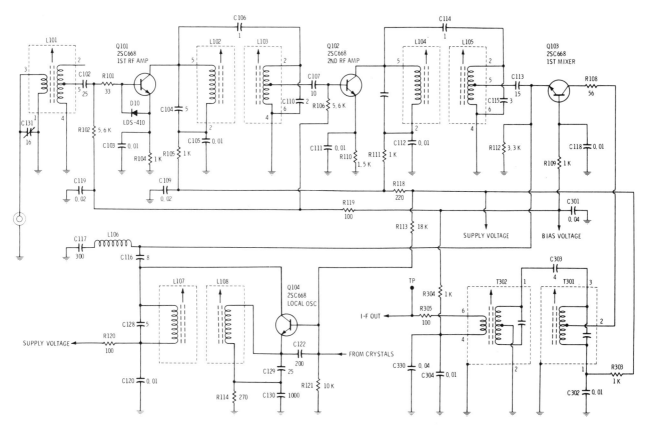

Courtesy Teaberry Electronics Corp.

Fig. 3-6. Two-stage rf amplifier configuration.

Scanner-Monitor Servicing Guide

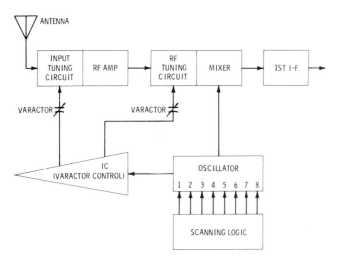

Fig. 3-7. Block diagram for an automatic rf tuning system.

ning process by varactors (voltage-variable capacitors) which are biased by dc output voltages from the IC. A varactor is a reverse-biased junction diode, whose junction capacitance depends upon the value of reverse-bias voltage that is applied. The value of reverse-biased voltage is determined by the oscillator crystal frequency at any given step in the scanning process.

Coil slugs and potentiometers are provided in the IC circuitry (not shown in Fig. 3-7) for precise adjustment of reverse-bias voltage on the varactors. However, only technicians who are familiar with the multifrequency signal-generation systems, add-on test modules, and output indicators should attempt to service the automatic tuning system. Technicians who are familiar only with fixed-tuned rf systems should send scanner-monitor receivers with automatic tuning systems to the manufacturer for alignment of the rf circuitry. Note that realignment is *not* required when oscillator crystals are changed in an automatic tuning system. When a new oscillator frequency is employed, the automatic tuning system will convert this new frequency into the correct value of bias voltage for the varactors, provided, of course, that the automatic tuning system is operating normally. Note that two-band and three-band receivers usually have the rf sections feeding into the same i-f input terminal, as shown in Fig. 3-8.

OSCILLATOR CIRCUITRY

A comparatively simple crystal-oscillator circuit for a manually selected monitor is shown in Fig. 3-9. Five quartz crystals are provided, any one of which can be selected by switch SW-1. Each crystal has a different frequency, although their frequencies should not be outside of the tuned-circuit coverage (for example, in the signal range from 37 to 43 MHz in Fig. 3-4). Each of the crystals is ground to a frequency which is 10.7 MHz above or below the particular channel frequency. In case image interference is encountered with the oscillator operating on the high side, for example, a crystal can be substituted that operates on the

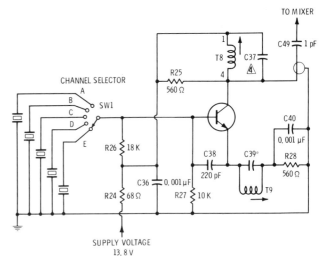

Fig. 3-9. Crystal-oscillator circuitry for a manually selected monitor.

low side of the channel frequency. Adjustment of the crystal-oscillator tuned circuits in Fig. 3-9 is made by connecting a vtvm or tvm to the mixer output lead and adjusting coils T8 and T9 for maximum output voltage. A reading of 0.04-volt peak is typical. Therefore, a high-sensitivity voltmeter must be used in this procedure.

Most scanner monitors employ electronic scanning, with diode-switch circuitry such as that shown in Fig. 3-10. When diodes D1 through D8 are reverse-biased, the crystals are not connected

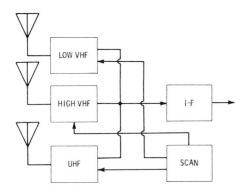

Fig. 3-8. Outputs from the three rf sections feed into the same i-f input terminal.

Tuning Circuits in Scanner-Monitor Receivers

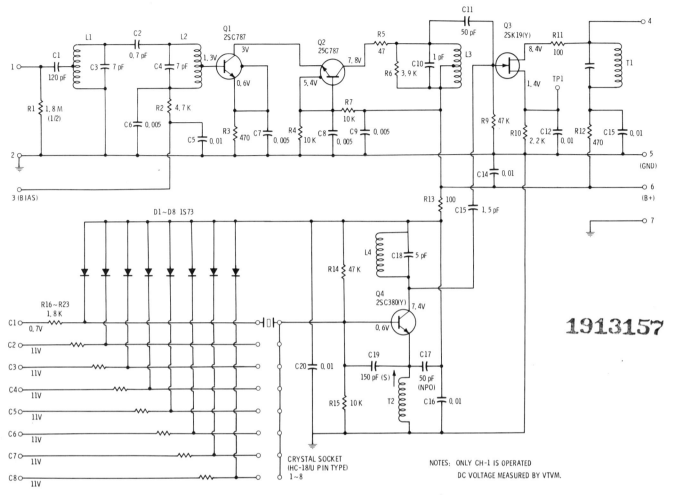

Fig. 3-10. Oscillator section with diode switching configuration.

Courtesy Radio Shack

into the collector circuit of oscillator transistor Q4. On the other hand, if a diode is forward-biased, the corresponding crystal is connected into the collector circuit of Q4. In turn, the oscillator operates at the frequency of that crystal. Bias voltages from the scanning logic are applied at terminals C1 through C8. In this example, the bias voltage is either 0.7 volt or 11 volts. The collector of Q4 operates at 7.4 volts, which also appears at the anode of each diode. Therefore, a bias voltage of 0.7 volt provides a forward bias on the given diode, whereas a bias voltage of 11 volts provides a reverse bias. Thus, in the example of Fig. 3-10, the C1 line (Channel 1) is activated, while the C2 through C8 lines are deactivated. Note in passing that a one-stage rf amplifier is utilized in this example—since Q2 is direct-coupled to Q1, these transistors operate in the same stage.

Many scanner monitors are designed as double-conversion, superheterodyne receivers. As depicted in Fig. 3-11, the first oscillator and the mixer convert the rf signal frequency to 10.7 MHz, after which the second oscillator and mixer convert the 10.7-MHz first i-f frequency to 455 kHz. Double conversion is employed for improved selectivity. A typical second oscillator mixer configuration is shown in Fig. 3-12. The second oscillator utilizes a crystal frequency of 11.155 MHz; the difference between this frequency and 10.7 MHz is 455 kHz. The second-oscillator frequency is always the same, regardless of the crystal frequency being used in the first oscillator. Diode CR3 in Fig. 3-12 operates as a voltage regulator to ensure that

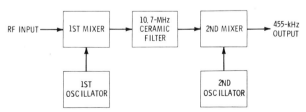

Fig. 3-11. Basic double-conversion arrangement.

Scanner-Monitor Servicing Guide

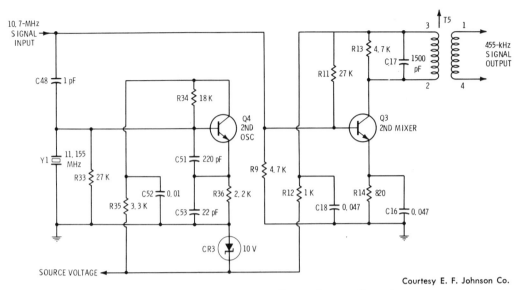

Fig. 3-12. Typical second oscillator-mixer configuration.

the second-oscillator frequency remains stable in case of a variation in supply voltage.

The uhf oscillators may supply either the third harmonic (overtone) of the crystal to the mixer, or the ninth harmonic. High-band vhf oscillators commonly supply the third harmonic of the crystal to the mixer. Low-band vhf oscillators may supply either the fundamental or the third harmonic of the crystal to the mixer. As an illustration, a typical scanner monitor has the following crystal frequency specifications:

Frequency for *low-band vhf, 30–50 MHz:*

Crystal fundamental frequency = channel frequency + 10.7 MHz

Frequency for *high-band vhf, 150–174 MHz:*

Crystal third overtone frequency = $\dfrac{\text{channel frequency} - 10.7 \text{ MHz}}{3}$

Frequency for *uhf band, 450–470 MHz:*

Crystal ninth overtone frequency = $\dfrac{\text{channel frequency} - 10.7 \text{ MHz}}{9}$

Fig. 3-13 shows how a frequency tripler follows the first oscillator in a typical high-band vhf front end. Since the third harmonic (overtone) of a crystal is comparatively weak, a transistor is gen-

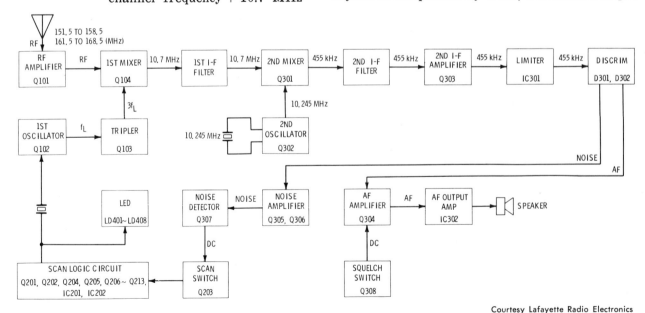

Fig. 3-13. A tripler stage follows the oscillator in this arrangement.

Tuning Circuits in Scanner-Monitor Receivers

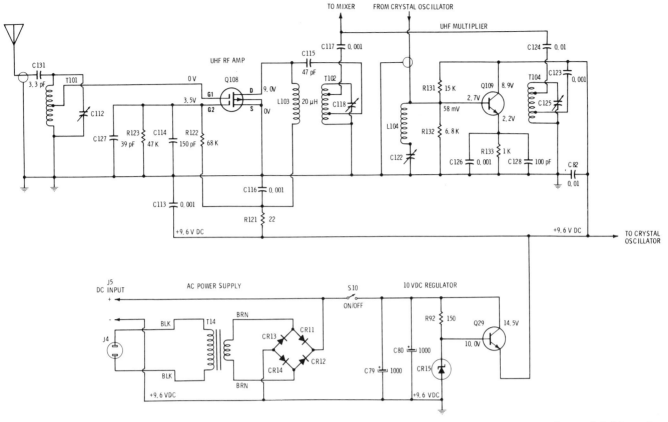

Fig. 3-14. Typical frequency tripler arrangement.

erally used as an amplifier in a frequency-multiplier stage. The tuned circuits in the multiplier stage will be resonated at the third harmonic of the crystal frequency, in this example. Fig. 3-14 exemplifies a tripler configuration in a uhf front end. To ensure that the frequency injected into the mixer is highly stable, a 10-volt dc regulator comprising Q29 and CR15 is utilized both for the multiplier stage and the crystal-oscillator circuit. When trouble occurs due to frequency drift, the dc voltages in the multiplier and crystal-oscillator circuits should be measured first, by using a high-accuracy voltmeter. Note that a ×9 multiplier section usually consists of two triplers connected in series.

Elaborate scanner-monitor receivers sometimes include automatic frequency control (afc) in the oscillator section, as depicted in Fig. 3-15. A typical configuration is shown in Fig. 3-16. The average voltage from the discriminator is normally zero when the oscillator is on frequency. On the other hand, if the crystal-oscillator frequency starts to drift, the average voltage will measure either positive or negative. In turn, this error voltage is amplified by Q101, Q102, and Q103 and is then applied across varactor diode CR101. Accordingly, the effective capacitance of CR101 will change, and this increase or decrease in capacitance will decrease or increase the oscillator frequency slightly. Thus, the oscillator is pulled back on-frequency. When afc trouble occurs, the fm discriminator might be more or less out of alignment, or there may be a defective component in the afc section. Direct-current voltage measurements are of basic importance in troubleshooting the afc section.

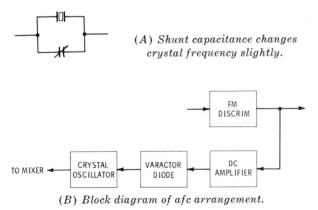

(A) *Shunt capacitance changes crystal frequency slightly.*

(B) *Block diagram of afc arrangement.*

Fig. 3-15. Automatic frequency control of crystal oscillator.

Scanner-Monitor Servicing Guide

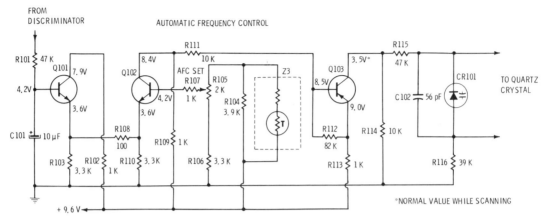

Fig. 3-16. Typical oscillator afc configuration.

RF ALIGNMENT PROCEDURES

Some scanner-monitor alignment procedures do not require highly accurate signal generators. However, alignment procedures for other receivers require very accurate frequency facilities. A service-type signal generator can be set to precise frequencies if it is calibrated against a crystal frequency standard such as that illustrated in Fig. 3-17. Another method of precise frequency measurement is to employ a frequency counter, such as the one shown in Fig. 3-18. Frequency counters are also used to adjust the oscillator frequencies in scanner monitors. Note that oscillator frequency measurements in uhf scanner monitors

Fig. 3-17. A crystal calibrator for accurate checks of generator frequencies.

Fig. 3-18. A typical digital-display frequency counter.

Tuning Circuits in Scanner-Monitor Receivers

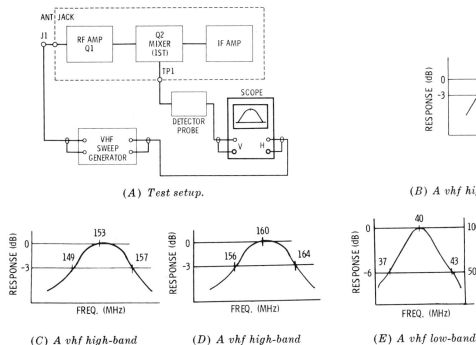

(A) Test setup.

(B) A vhf high-band response curve.

(C) A vhf high-band response curve with less bandwidth.

(D) A vhf high-band response curve with a different peak frequency.

(E) A vhf low-band response curve with −6-db points specified.

(F) A vhf low-band response curve with a different peak frequency.

Fig. 3-19. Sweep alignment of a scanner-monitor rf section.

are made at ⅓ or ⅑ of the mixer injection frequency. On the other hand, when frequency measurements are specified in the multiplier circuits of a uhf scanner, the frequency counter will need a range up to 500 MHz.

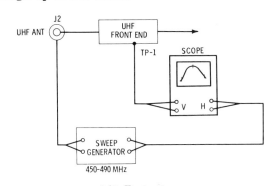

(A) Test setup.

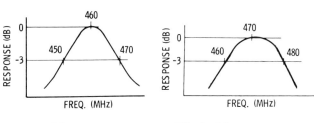

(B) A uhf response curve.

(C) A uhf response curve with a different peak frequency.

Fig. 3-20. Sweep alignment of a uhf scanner-monitor rf section.

An rf section can be aligned with a standard signal generator or with a sweep generator. When a sweep generator is used, a marker generator is also required. If the sweep generator does not have built-in marker facilities, a signal generator can be used to mark the response curve. An rf alignment setup and specified response curve are shown in Fig. 3-19. A uhf alignment setup is basically the same, except that a uhf sweep generator is used, as shown in Fig. 3-20. Note that a detector probe is indicated in Fig. 3-19, whereas a direct connection to the scope is indicated in Fig. 3-20. This requirement depends upon the test point specified by the manufacturer; the scanner-monitor servicing data should always be consulted in this regard.

As an example of rf alignment adjustments, the cores of L101, L102, L103, L104, and L105 (Fig. 3-6) would be adjusted as required to obtain the desired peak frequency and the specified bandwidth. As another example, trimmer capacitors C112 and C118 (Fig. 3-14) would be adjusted as required to obtain the chosen peak frequency and the specified bandwidth. Next, the oscillator frequency is checked as shown in Fig. 2-21. The servicing data may specify either inductive coupling to a pickup loop, or capacitive coupling through a small fixed capacitor. The crystal frequency depends to a slight extent upon the capacitance and

Scanner-Monitor Servicing Guide

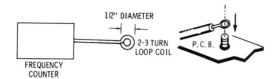

(A) *Pickup coil.*

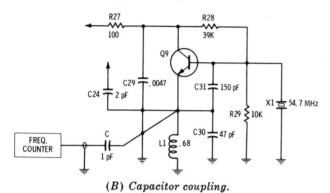

(B) *Capacitor coupling.*

Fig. 3-21. Frequency-counter coupling to oscillator coil.

inductance values in the oscillator circuit. For example, the crystal frequency in Fig. 3-6 can be changed slightly by adjustment of L107 and/or L108. With reference to Fig. 3-9, the crystal frequency is slightly affected by adjustment of T8 and/or T9.

TROUBLESHOOTING PROCEDURES

1. Low Sensitivity

Probable causes of low sensitivity due to defects or malfunctions in the rf or oscillator sections of a scanner monitor are as follows:

a. An oscillator crystal has been inserted that has a frequency at one end of the range, whereas the rf circuits have been peaked at the other end of the range. Check alignment of rf section.
b. The rf circuits have been aligned with inaccurate generators. Check the calibration of the signal generator or marker generator.
c. Defective rf amplifier transistor. Measure transistor terminal voltages.
d. Deteriorated oscillator transistor. Measure transistor terminal voltages.
e. The rf coil may have been replaced with an unsuitable type. Use replacement coil specified in the servicing data.
f. Oscillator crystal may be operating slightly off-frequency. Check operating frequency with a frequency counter.
g. If a FET has been replaced in the rf section, the shorting spring may not have been removed from the FET. After all of the FET leads have been soldered into the printed-circuit board, be sure to remove the shorting spring.

2. Signal Channel Is Dead

Typical causes of a dead signal channel due to defects or incorrect operation in the rf or oscillator sections are as follows:

a. Oscillator crystal has been inserted into wrong socket.
b. Oscillator section has been incorrectly programmed.
c. Defective antenna connector.
d. Defective component in afc section. Measure dc voltages in the afc circuits. Also, check afc alignment according to the servicing data.
e. Oscillator tank coil resonates outside permissible range. Align oscillator tuned circuits as described in the servicing data.
f. Oscillator crystal defective (not likely, but possible).

3. Intermittent Reception

Possible causes of intermittent operation due to defects in the rf or oscillator sections are as follows:

a. Defective crystal socket, or crystal making poor contact with socket springs. Note that phosphor-bronze springs will break if bent excessively.
b. Oscillator supply voltage marginal. Measure oscillator voltage.
c. Oscillator tank tuned to "ragged edge" of permissible range. Align according to the servicing data.
d. Intermittent oscillator transistor. Tap the transistor; check its response to moderate heat.
e. Microscopic break in printed-circuit conductor; inspect under magnifying glass; flex board slightly while making continuity tests.
f. Cold-soldered connection in rf or oscillator circuitry. Reheat any suspicious joints; use heat sinks on semiconductor leads while reheating associated connections.

4. Volume Change Excessively During Reception

Common causes for excessive variation in volume during reception are as follows:

a. Mobile transmitter is traveling along steel structures, below underpasses, or past wooded areas. Compare reception on another channel.

b. Battery connector may be loose or corroded; check supply voltage with voltmeter.
c. Volume control may be worn and erratic; check with ohmmeter.
d. Oscillator crystal drifting; make substitution test, or check with frequency counter.
e. Because of the high sensitivity of these receivers, "skip" signals may be received at times by scanner monitors with automatic tuning systems. This kind of signal is likely to fluctuate in and out of audibility.

5. Noisy Reception

Possible causes of noisy reception due to defects or malfunctions in the rf and oscillator sections are as follows:

a. A scanner monitor that has an automatic tuning system and is used with a high-gain antenna is likely to develop noise and/or interference. Try using a short or low-gain antenna.
b. Excessive noise may be caused by a deteriorating mixer transistor. Connect a bypass capacitor temporarily across the i-f input terminals to determine whether it stops the noise. (Open squelch for test.)
c. Noise is occasionally introduced by an external power supply. Cross-check with battery supply.
d. Nearby fluorescent lamps can cause noisy reception. Check operation in another location.
e. Off-on switches occasionally become worn and noisy; try connecting a jumper across the switch terminals.

CHAPTER 4

I-F Circuitry and Trouble Analysis

The i-f sections in most scanner monitors operate at a center frequency of 10.7 MHz. Double conversion is often used, with center frequencies of 10.7 MHz and 455 kHz. The receiver service data should be consulted, in any case. For example, Fig. 4-1 shows the specified response curves for the i-f amplifier and fm detector of a scanner monitor that operates at a center frequency of 10.8 MHz. The i-f bandwidths are quite narrow, compared to rf bandwidths. As an illustration, the bandwidth specified in Fig. 4-1 is slightly less than 20 kHz. Fig. 4-2 shows an S curve with a total frequency excursion of 16 kHz. The normal amplitude of the curve is 0.4 volt p-p.

Frequency-response curves such as that shown in Fig. 4-1 are usually obtained by means of conventional tuned circuits and ceramic bandpass filters in the i-f section of a scanner-monitor receiver. Fig. 4-3 shows the appearance of a ceramic filter. A few scanner monitors use quartz-crystal filters in addition to ceramic filters, and an occasional design utilizes LC bandpass circuitry only. Ceramic filters contain barium titanate crystals which are very stable unless subjected to mechanical damage. This type of crystal can also withstand comparatively high temperatures. The fm detectors in scanner monitors usually employ conventional tuned circuits; however, an occasional design uses a triplet of ceramic resonators. When a ceramic filter is replaced, it is good practice to check the i-f response curve and to make touch-up adjustments of the tuned circuits if necessary.

TYPICAL I-F CIRCUITRY

Some i-f amplifiers, such as the one shown in Fig. 4-4, use discrete transistors. This is an example of a double-conversion configuration that employs four i-f stages. The mixer transistors also provide some conversion gain. An fm slope detector is utilized, which operates with ceramic resonator Z3 instead of a conventional tuned circuit. Note that the slope detector is preceded by limiter diodes CR5 and CR6. A slope detector changes a

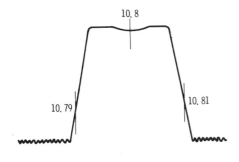
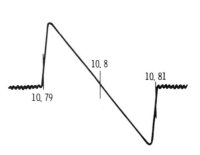

Fig. 4-1. Specified response curves for the i-f amplifier and fm detector of a monitor receiver.

Scanner-Monitor Servicing Guide

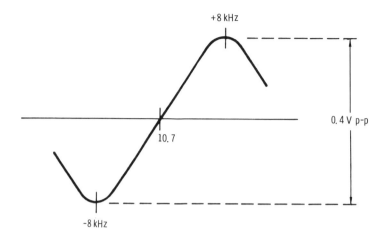

Fig. 4-2. A discriminator S curve with a 16-kHz bandwidth.

Fig. 4-3. Appearance of a ceramic filter in an i-f strip.

frequency-modulated waveform into an amplitude-modulated waveform. Therefore, an a-m detector must follow the slope detector. Diode CR7 operates as an a-m detector. The first-mixer coupling coil, L2, is peaked at 10.7 MHz. The second-mixer coupling transformer, T2, is peaked at 455 kHz. Maximum 10.7-MHz conversion gain is obtained when the first oscillator is adjusted to inject 34 mV rms into the base of Q2. In normal operation, the second oscillator injects 31 mV rms into the base of Q6.

Tuned circuits are not used to couple between Q7, Q8, Q9, and Z3 in Fig. 4-4. Instead, RC coupling is used, via capacitors C28, C29, and C30. Normal terminal voltages are specified for the transistors with the scanner monitor squelched and with no rf signal input. The gain of the rf and i-f systems is checked by means of a quieting-sensitivity test. In this example, an ac vtvm or tvm is connected at the output of the audio amplifier. A lab-type rf signal generator, set for a 1-microvolt signal at the center frequency of the selected channel, is connected to the antenna input terminal. With the scanner monitor unsquelched, the meter reading should drop 20 dB when the generator signal is applied. In this example, if it takes more

I-F Circuitry and Trouble Analysis

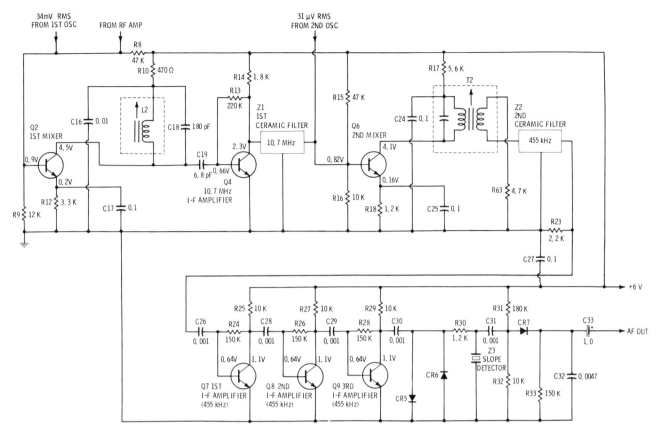

Fig. 4-4. An i-f amplifier configuration that uses discrete transistors.

than 5.5 microvolts for 20-dB quieting on the "worst channel," the technician would conclude that the sensitivity of the signal channel is subnormal. This test is based on the ability of a weak signal to saturate the limiter and thereby reduce the a-m noise amplitude.

Other i-f amplifiers, such as that shown in Fig. 4-5, use integrated circuits and transistors. In this example, transistor Q101 operates as a 10.7-MHz amplifier. Integrated circuit IC101 operates as a 10.7-MHz amplifier, second oscillator, and second mixer with a 455-kHz output. Crystal Y101 determines the second-oscillator frequency. In turn, ceramic filter CF-1 provides the required i-f selectivity. Integrated circuit IC102 functions as an i-f amplifier and as a quadrature fm detector. The quadrature coil is L103, which is adjusted to obtain the specified S curve. The i-f alignment is made by adjustment of transformers T101 and T102. Fig. 4-6 shows the layout of the circuit board corresponding to the configuration in Fig. 4-5. There are two basic reasons for the seemingly random placement of devices and components on a circuit board. First, the printed-circuit conductors must be routed so that they do not cross; second, high-frequency conductors must be kept as short as possible. Another consideration is placement of input and output components, such as transformer T101 and coil L103, in locations that will minimize stray coupling.

Note in Fig. 4-6 that a dot is shown at one end of IC101 and IC102. These dots correspond to dots on the IC package and serve as a guide in replacement of the integrated circuits. In other words, if an IC is reversed end-for-end, the i-f amplifier will be inoperative. A scanner-monitor receiver may use a ceramic filter that is not a package unit but is arranged from discrete components including two ceramic resonators, as shown in Fig. 4-7. Again, a series of capacitively coupled tuned circuits may be used instead of a ceramic filter, as exemplified in Fig. 4-8. The coupling capacitors in this arrangement have values that provide correct filter bandwidth when all of the coils are peaked at 455 kHz.

An elaborate RC-coupled i-f amplifier configuration for a scanner monitor is seen in Fig. 4-9. Note that only the input, output, and mixer-load circuits are tuned. The input circuit consists of a 10.7-MHz ceramic filter, and the output circuit

Scanner-Monitor Servicing Guide

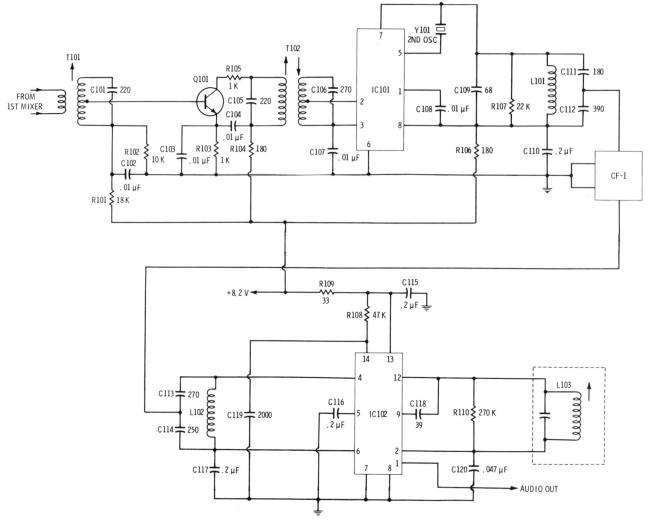

Fig. 4-5. An i-f configuration that uses both ICs and a transistor.

Courtesy Regency Electronics, Inc.

comprises a discriminator arrangement. Nine coupled transistors are used between the input and output circuits. Transistor Q1 is a 10.7-MHz amplifier; Q2 is an oscillator that feeds into Q3, which operates as a 455-kHz mixer. Observe that the mixer section is untuned, except for the 455-kHz ceramic filter between Q3 and Q4. There are three stages of i-f amplification, with Q4/Q5, Q6/Q7, and Q8/Q9 in a cascade or "stacked" arrangement. Thus, the collector of Q4 drives the base of Q5, the collector of Q6 drives the base of Q7, and so on. Although more transistors are employed in this design than in other i-f strips, the sensitivity is approximately the same, due to the absence of tuned coupling circuits.

Top and bottom views of the printed-circuit board for the configuration of Fig. 4-9 are shown in Fig. 4-10. Note that transistors Q1, Q2, and Q3 appear somewhat "out of sequence" in the layout, although Q4, Q5, Q6, Q7, Q8, and Q9 are placed in the same order that they appear in the schematic diagram. In other words, it cannot be assumed that transistors will be located in the order that would be expected from looking at the circuit diagram. The ceramic filters are placed in "logical positions" with respect to the transistors on the circuit board. Similarly, discriminator coils T1 and T2 are found in anticipated locations. On the other hand, capacitors C8, C12, and C16 are mounted in locations different from those that might be expected from inspection of the schematic diagram. In other words, the technician must occasionally check over a considerable area of the circuit board in order to locate a particular component corresponding to its symbolic identification in the schematic.

The dc voltage measurements are not always indicated on schematic diagrams. Instead, the volt-

I-F Circuitry and Trouble Analysis

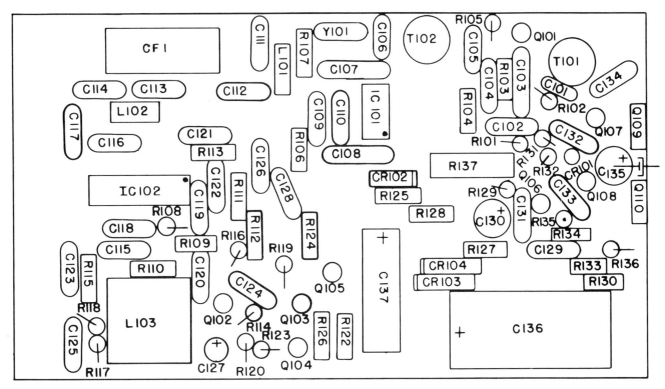

Courtesy Regency Electronics, Inc.

Fig. 4-6. Circuit board layout for the configuration in Fig. 4-5.

ages may be specified in a separate table. For example, Fig. 4-11 shows the circuit diagram for an i-f amplifier that uses two integrated circuits and no discrete transistors. Each IC has terminals that are closely spaced. In turn, the diagram would

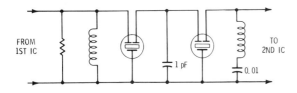

Fig. 4-7. A ceramic filter having discrete components including two ceramic resonators.

become comparatively cluttered and confusing to read if the dc voltages were indicated at the IC terminals. Therefore, a separate voltage table is provided in this example, as shown in Fig. 4-11B. Occasionally, a scanner monitor may be encountered for which no voltage data are available. In such a case, it is sometimes possible to obtain a similar scanner monitor that is in normal operating condition and to make comparative voltage measurements. Note also that comparative point-to-point resistance measurements and resistance-to-ground measurements may be made. It is advisable to use a hi-lo ohmmeter (Fig. 4-12) for these types of resistance measurements.

SEMICONDUCTOR TESTING

Measurement of dc voltages at transistor terminals is the basic in-circuit method of checking for defects. For example, with reference to Fig. 4-4, a voltage of less than 2.3 volts at the collector of Q4 would indicate that a fault is present. If the collector voltage were found to be 1.5 volts, for example, the trouble could be due to several things: a leaky collector junction in Q4, leakage in capacitor C19, an increase in the resistance of R14, or a decrease in the resistance of R13. Therefore, further tests are required in this situation to pinpoint the defect. A very useful test in this case is to open the connection to C19. This can easily be done by using a razor blade to cut a slit across the printed-circuit conductor at the end of C19. Now, if the

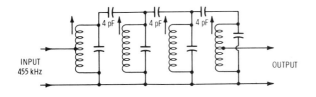

Fig. 4-8. An i-f LC bandpass filter configuration.

Scanner-Monitor Servicing Guide

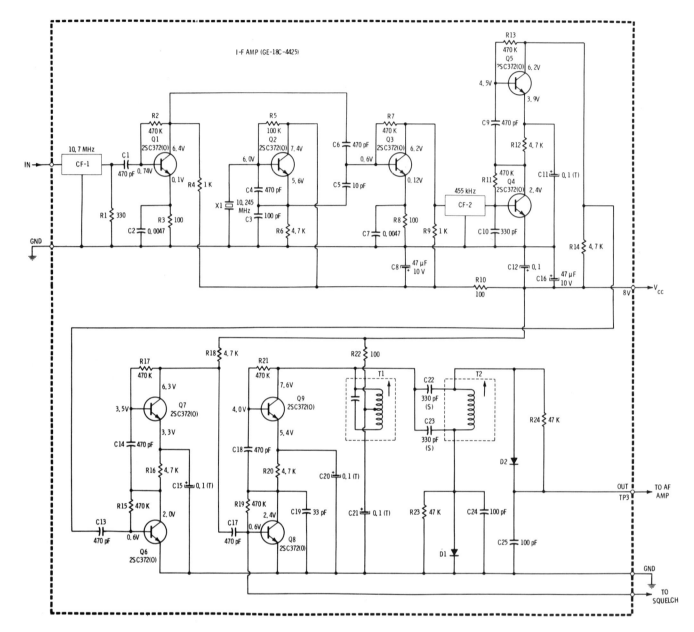

Courtesy Radio Shack

Fig. 4-9. An elaborate RC-coupled i-f system.

collector voltage on Q4 measures 2.3 volts ±10%, the conclusion is that C19 is leaky and must be replaced. The printed-circuit foil is repaired by melting a small drop of solder across the cut.

Next, suppose that after C19 is disconnected from the base of Q4 in Fig. 4-4, the collector voltage still measures 1.5 volts. The reasonable conclusion in this case is that the collector junction of Q4 is leaky. In other words, transistors are more likely to deteriorate than resistors. Many technicians would replace Q4 at this time and repair the slit in the printed-circuit conductor. However,

since it is somewhat tedious to replace a transistor, other technicians will make additional tests at this time. For example, the collector of Q4 may be disconnected from its circuit by cutting a slit across the foil conductor, and the base of Q4 may be similarly disconnected from R13. In turn, R13 and R14 can be measured with an ordinary ohmmeter. In case these resistors are found to have correct values, replacement of Q4 is fully justified. After the transistor is replaced, the slits in the printed-circuit conductors are repaired with small drops of solder.

I-F Circuitry and Trouble Analysis

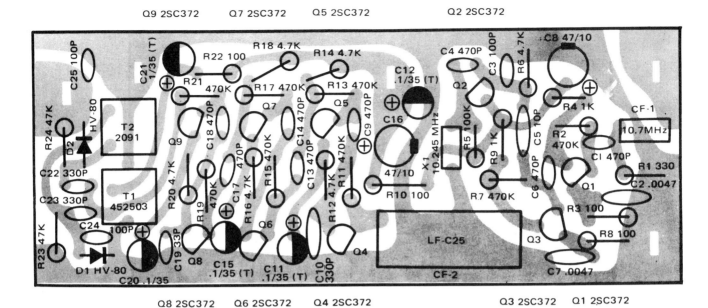

(A) Top view.

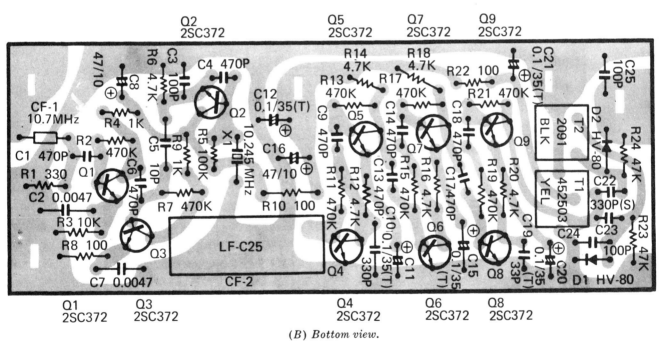

(B) Bottom view.

Courtesy Radio Shack

Fig. 4-10. Views of printed-circuit board for i-f configuration in Fig. 4-9.

Note in passing that a hi-lo multimeter such as the one illustrated in Fig. 4-12 can be used to measure the values of resistors in-circuit, provided that the transistor is normal. The lo-power ohmmeter applies less than 0.08 volt across the resistor under test, which is insufficient voltage to turn on a transistor junction. In other words, a lo-power ohmmeter could be used to measure the values of R14 and R13 in-circuit (Fig. 4-4), provided that Q4 were normal. This is why a lo-power ohmmeter is useful to make various kinds of in-circuit resistance measurements. In various situations, it is also helpful to check the conditions of transistors in-circuit. This is generally practical by means of an in-circuit transistor tester, such as the one illustrated in Fig. 4-13. In case doubt

49

Scanner-Monitor Servicing Guide

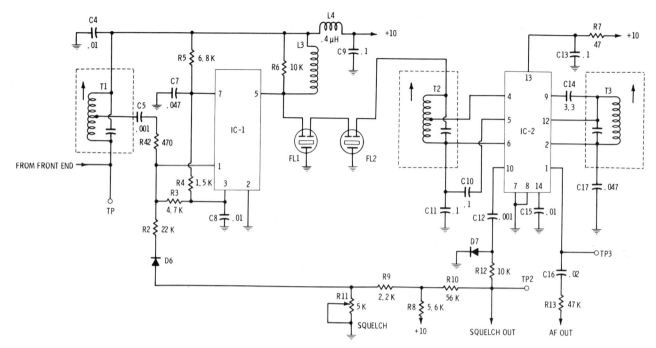

(A) Schematic diagram.

PIN NO.	I.C. NUMBER						
	1	2	3	4	5	6	7
1	1.3	4.9	11	NC	NC	5.0	.13
2	GND	3.5	NC	4.4	5.0	5.0	.13
3	1.4	NC	7.8	5.0	.13	9.0	9.0
4	NC	1.4	.7	1.5	NC	5.0	.13
5	10	1.4	.5	.1	.13	.1	5.0
6	NC	1.4	NC	4.4	5.0	9.0	9.0
7	2.1	GND	0	GND	GND	GND	GND
8	NC	GND	GND	5.0	5.0	9.0	.2
9		.14	NC	.1	.13	.1	5.0
10		1.4	GND	NC	1.4	.1	5.0
11		NC	.5	6.2	5.0	9.0	9.0
12		3.5	6.5	5.0	5.0	.1	5.0
13		10	0	NC	1.0	5.0	.13
14		5.3	12	6.2	6.2	.1	4.4

VOLTAGES ARE MEASURED WITH "MANUAL-SCAN" SWITCH IN MANUAL POSITION, "VOLUME" CONTROL COUNTERCLOCKWISE AND "SQUELCH" COUNTERCLOCKWISE OR, FOR SOME MEASUREMENTS CW/CCW.

(B) Table of dc voltages.

Courtesy Electra Co.

Fig. 4-11. An i-f configuration using 2 ICs.

should occur, the printed-circuit conductors can be slit on two of the transistor terminals for an out-of-circuit crosscheck.

It is often possible to make useful in-circuit transistor tests without the use of specialized equipment. For example, Fig. 4-14 shows how a transistor turn-off test can be made in a typical situation. A voltmeter is applied between the collector and emitter terminals of the transistor. In turn, a temporary short circuit is applied between the base and emitter terminals. If the transistor has normal control action, the meter reading will jump to 6 volts, the value of the supply voltage. In other words, when the base and emitter terminals are short-circuited, the transistor is zero-biased. Next, observe that when the base is biased from the collector, as is Q4 in Fig. 4-4, the collector will not rise fully to the supply voltage in a turn-off test, even if the transistor is normal. For example, in this situation, the collector will be connected to the emitter through a 220K resistor during the test, and the collector voltage will rise to approximately 5.94 volts, if the transistor in this circuit is normal.

I-F Circuitry and Trouble Analysis

Another basic in-circuit transistor test is depicted in Fig. 4-15. In this test, called a turn-on test, a voltmeter is applied between the collector and emitter terminals of the transistor. Then, a temporary 50K bleeder resistor is applied between the base of the transistor and the collector supply voltage, V_{cc}. If the transistor has control action, the voltmeter reading will fall when the bleeder resistor is applied. This test is based upon the increased collector current flow that normally occurs when the forward bias of the transistor is increased. An increase in collector current produces a greater voltage drop across load resistor R_L, with the result that the collector-to-emitter voltage normally decreases.

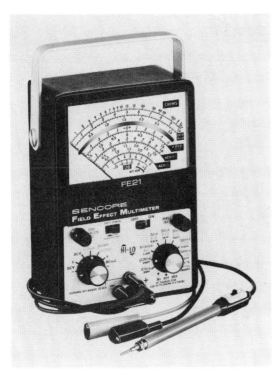

Courtesy Sencore

Fig. 4-12. A hi-lo field-effect multimeter.

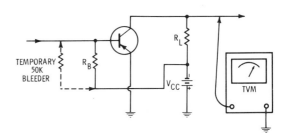

Fig. 4-15. Transistor turn-on test.

Transistor terminal voltages are usually measured with respect to ground. However, it is sometimes difficult to measure bias voltages accurately with respect to ground, because the bias voltage may be equal to a small difference between two comparatively large voltage values, as exemplified in Fig. 4-16. Therefore, it is good practice to measure the bias voltage directly between the base and

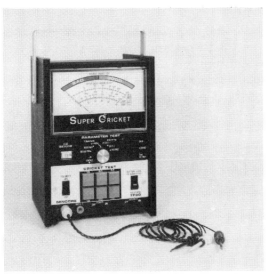

Courtesy Sencore

Fig. 4-13. An in-circuit transistor tester.

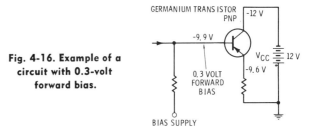

Fig. 4-16. Example of a circuit with 0.3-volt forward bias.

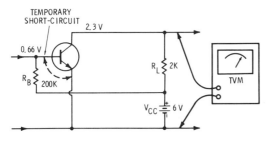

Fig. 4-14. Transistor turn-off test.

emitter terminals in this situation. It is also quicker to measure the bias voltage from base to emitter because a single measurement suffices and no subtraction is required. Some apprentice technicians tend to become confused about polarity relationships in this kind of circuit when an emitter reference point is used. If an emitter reference point is used instead of a ground reference point,

51

as shown in Fig. 4-17, it might seem that the transistor terminal voltages have changed from positive to negative. Of course, the voltage distribution is basically the same in either case, and both circuits provide a forward bias of 0.3 volt on the transistors. In the case of Fig. 4-17B, the base is negative with respect to the emitter potential of zero volts. Accordingly, the emitter is positive with respect to the base potential of −0.3 volt. Thus, the transistor is forward-biased by 0.3 volt.

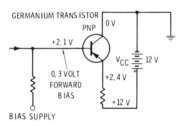

(A) Ground reference point.

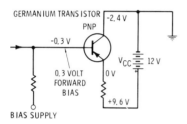

(B) Emitter reference point.

Fig. 4-17. Polarity of measured voltage depends upon reference point.

Next, consider the turn-off test for a transistor with a grounded collector circuit, as shown in Fig. 4-18. To understand the test action, it is helpful to simplify the configuration into its equivalent dc circuit. When a temporary short circuit is applied between the base and emitter of the transistor, no collector current normally flows. In turn, the voltmeter reads zero volts across the collector dropping resistor. On the other hand, if the transistor is leaky, more or less collector current will continue to flow when the base and emitter are shorted together, and a voltage reading will then be obtained across the collector dropping resistor. Note that there are some configurations in which it is impractical to make a turn-off test. For example, if there were no collector dropping resistor present in the circuit of Fig. 4-18, a turn-off test could not be made.

When a turn-off test cannot be made, it is often practical to make a turn-on test, as shown in Fig. 4-19. In this situation, the voltmeter is connected across the emitter resistor, and a 50K bleeder resistor is temporarily applied between the transis-

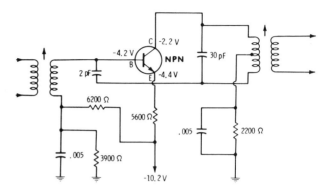

(A) With collector dropping resistor.

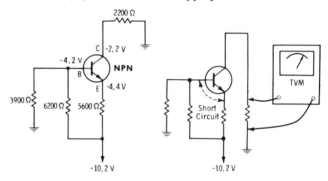

(B) Equivalent dc circuit.　　(C) Control-action test.

Fig. 4-18. Turn-off test in stage with grounded collector circuit.

tor base and the collector. In turn, the forward bias on the transistor is increased, and the collector (emitter) current normally increases. This current increase causes a greater voltage drop across the emitter resistor, provided that the transistor has normal control action. Turn-on and turn-off tests are particularly useful when you are troubleshooting scanner-monitor circuitry in which normal dc voltages are not specified.

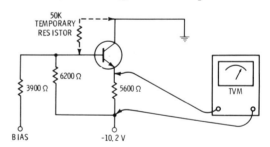

Fig. 4-19. Turn-on test in a grounded-collector circuit.

TROUBLESHOOTING PROCEDURES

1. Low Sensitivity

Probable causes of low sensitivity due to defects or malfunctions in the i-f section are as follows:

a. Leaky or open coupling capacitor, such as C14 in Fig. 4-11.

I-F Circuitry and Trouble Analysis

b. Leaky or open bypass capacitor, such as C8 in Fig. 4-11.
c. Defective transistor such as Q1 in Fig. 4-9 (noise level high).
d. Defective transistor such as Q9 in Fig. 4-9 (noise level low).
e. Misaligned tuned i-f circuit, such as T102 or L103 in Fig. 4-5.
f. Defective integrated circuit, such as IC101 in Fig. 4-5.
g. Faulty ceramic filter, such as Z1 in Fig. 4-4 (not likely, but possible).

2. Does Not Scan and Squelch Does Not Operate

Typical causes of no scan or squelch action due to i-f defects are as follows:

a. Faulty squelch control.
b. Defective i-f component such as CR5, CR6, or both, in Fig. 4-4.
c. Malfunctioning integrated circuit, such as IC2 in Fig. 4-11.
d. Excessively noisy limiter transistor.
e. Incorrect bias voltage(s) in limiter circuit.

3. Noise Is Audible; No Audio Signal Output

Possible causes of audible noise but no audio signal output due to i-f troubles are as follows:

a. Malfunction in fm detector section, resulting in a-m detection only. The stabilizing capacitor in a ratio detector may be defective.
b. Open or short circuit in mixer output circuit, or i-f input circuit.
c. Operation near a high-level noise source that masks the audio signal. Check reception in another location.
d. Excessively noisy i-f transistor. Check output from each i-f stage will scope. (Scope must have adequate response at 10.7 MHz.)

4. Distorted Output

Common causes of distorted audio output due to defects or malfunctions in the i-f section are as follows:

a. Discriminator, ratio detector, or slope detector misaligned.
b. Defective component in fm detector, such as diode D2 in Fig. 4-9.
c. Regeneration or marginal oscillation in i-f circuits. Check bypass and decoupling capacitors.
d. Faulty IC that includes the fm detector section, such as IC2 in Fig. 4-11. Make substitution test.

e. Defective ceramic filter (not likely, but possible).

5. Intermittent Operation

Probable causes of intermittent operation due to defects or malfunctions in the i-f amplifier are as follows:

a. IC not fully inserted into socket, or corroded contacts.
b. Cold solder connection on printed-circuit board.
c. Intermittent transistor; tap transistor and monitor circuit voltages.
d. Intermittent capacitor; check response to heat and to tapping.
e. Intermittent resistor (not likely, but possible). Tap suspected resistors; check response to heat.

6. Interference in Reception

Typical causes of interference in reception due to faults in the i-f section are as follows:

a. Seriously misaligned bandpass filter, such as the configuration in Fig. 4-8. Note that open shunt capacitors can cause this condition.
b. Defective ceramic filter or shorted filter connections. Check i-f response curve; replace filter if it appears to be defective.
c. Open ground connection to ceramic filter. A "floating" filter may have an extremely broad response.
d. Operation of scanner-monitor receiver in location where i-f circuit picks up strong interference directly. Try another location.

7. Fading or Fluctuation in Audio Output Level

Possible causes of fading or fluctuation in audio output level due to defects in the i-f section are as follows:

a. Loose slug in tuned i-f coil shifts in position as scanner-monitor receiver is moved about.
b. Unstable transistor; monitor output at each i-f stage in succession with oscilloscope (scope must have adequate response at 10.7 MHz).
c. Injection voltage may be varying at first or second mixer; monitor with scope or high-frequency tvm.
d. Integrated circuit could be making a high-resistance contact to a socket terminal. Check contacts.
e. Unstable integrated circuit. Make substitution test of suspected unit.

CHAPTER 5

Scan Circuits and Indicators

Scan circuits are basically electronic switching arrangements that successively sample the channels in a scanner-monitor receiver to sense whether an incoming signal may be present. Each channel is sampled for approximately $\frac{1}{10}$ second. This scanning action continues until a signal appears on one of the channels. Thereupon, the scan action stops and the monitor locks onto the active channel. At the end of the transmission, the monitor will remain locked onto the channel for approximately two seconds before scanning is resumed. This delay ensures that the entire transmission will be received in case of a brief signal interruption. Lockout switches are usually provided in the scan network so that any desired channel(s) can be "skipped" during the automatic scanning operation. In addition, most scanners have provisions for manual scanning, and the monitor can be stepped one channel at a time by flipping a switch.

Both integrated circuits and transistors are used in scan circuits. As shown in Fig. 5-1, automatic scanning action is stopped and started by the squelch (electronic) switch. The scan circuit controls the first local oscillator by means of diode switches that connect one crystal at a time into the oscillator circuit. In addition, the scan circuit turns the appropriate indicator lamp on to show what channel is enabled at any time. Fig. 5-2 shows the configuration for a comparatively simple scanning section. The scanning rate is controlled by a multivibrator (scan clock) consisting

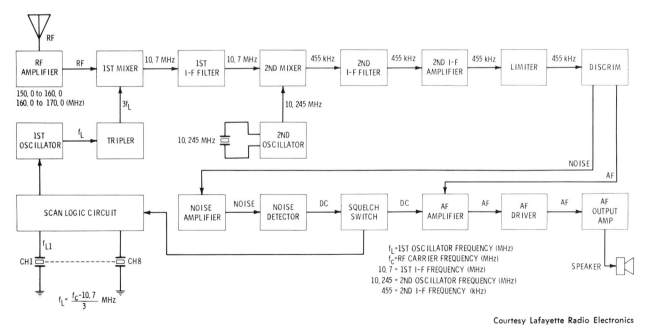

Courtesy Lafayette Radio Electronics

Fig. 5-1. Scan logic circuit operates between the crystals and the first oscillator.

Scanner-Monitor Servicing Guide

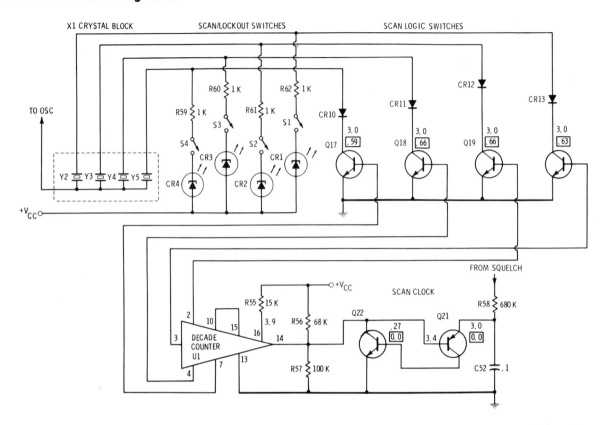

Fig. 5-2. Configuration for a four-crystal scan circuit.

Courtesy E. F. Johnson Co.

of Q21 and Q22. In turn, the output from the multivibrator is applied to decade counter U1. A conventional decade counter provides counts (output pulses) in succession up to 9 and then resets. In this particular application, U1 is connected to provide four output pulses in succession, and then reset. The counting sequence at the IC terminals is 3, 2, 4, 7; the counting sequence then repeats.

Output pulses from the decade counter are applied in succession to transistors Q17, Q18, Q19, and Q20 in Figure 5-2. Thus, only one transistor conducts at a time. Suppose that a positive pulse is applied to the base of Q20, causing Q20 to come out of cutoff and conduct. In turn, CR13 is forward-biased and also conducts. This provides a low-resistance path to ground for crystal Y2, and the oscillator will operate at the frequency of Y2 as long as the positive pulse is present on the base of Q20. Note that Y2 can be prevented from oscillating (locked out) by opening switch S1. In other words, the collector of Q20 is disconnected from the dc supply voltage if S1 is open. Observe also that collector current for Q20 flows through light-emitting diode CR1. Accordingly, CR1 will glow as long as crystal Y2 is connected into the oscillator circuit. The other three crystals and their corresponding LEDs operate in the same manner.

BASIC SCAN-CIRCUIT TESTS

A tvm and scope are the basic test instruments for checking scan circuits. With reference to Fig.

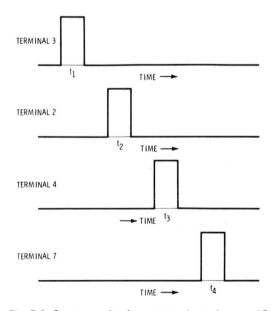

Fig. 5-3. Sequence of pulse outputs from the scan IC.

Scan Circuits and Indicators

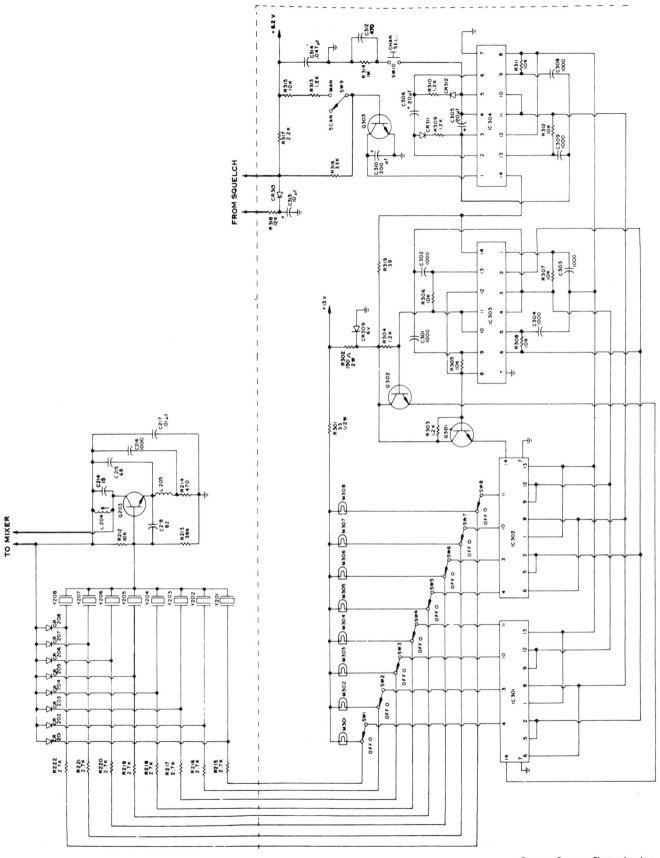

Fig. 5-4. Scan circuitry for an 8-channel oscillator.

5-2, dc voltages are specified at the collectors of the transistors. Two values are given. When a pulse output is present from decade counter U1, the collector of the associated transistor goes "logic low" to approximately 0.6 volt. On the other hand, when a pulse output is not present from U1, the collector voltage goes "logic high" to 3.0 volts. When an oscilloscope is applied at the output terminals of U1, pulse waveforms are normally displayed as shown in Fig. 5-3. These pulses have a normal amplitude of approximately 2.4 volts p-p. Note that pulse waveforms will be displayed only when the network is scanning. In case the monitor is stopped on a channel or if all of the lockout switches are open, no pulse waveforms are present.

Next, it is helpful to consider the scan circuitry for an 8-channel oscillator, shown in Fig. 5-4. This arrangement has two IC counters, IC303 and IC-304. These counters are also called 1-out-of-4 decoders; together, they form a 1-out-of-8 decoder. Integrated circuits IC301 and IC302 operate as indicator drivers. Small incandescent lamps are used as indicators in this example. The dc voltage distribution for the circuit in Fig. 5-5 depends upon whether the unit is scanning or whether it is being manually operated. Manual-operation voltages are specified for Channel 1. When the scanner section of the receiver is operating, some of the voltages are pulsating. In turn, a dc voltmeter will indicate less than the peak voltage of the pulse. This condition is shown in Table 5-1 by a "P" beside the specified voltage value. For example, the voltage at pin 14 of IC301 is specified at 3P volts while the scanner is operating, but is specified at 5.2 volts during manual operation with the monitor switched to channel 1.

CHANNEL INDICATORS

Small light-emitting diodes (LEDs) such as those used for channel indication in scanner monitors have a forward voltage drop of approximately 2 volts and a maximum permissible reverse voltage of 3 or 4 volts. Fig. 5-5A shows the appearance and terminal identifications of typical LEDs. The typical operating range for an LED is depicted in Fig. 5-5B, with current and voltage ratings for typical devices. In most cases, a current flow of 15 to 20 mA will provide sufficient illu-

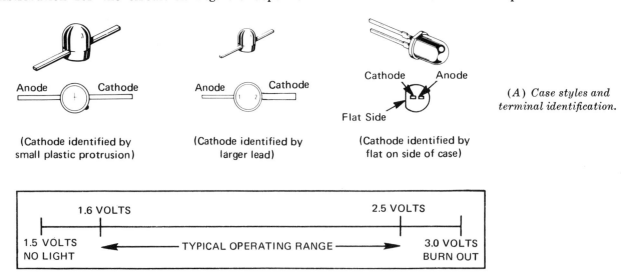

(A) Case styles and terminal identification.

(B) Operating characteristics.

Maximum Ratings

HEP Type No.	Reverse Voltage V_R (Volts)	Forward Current I_F (mA)	Power Dissipation P_D (mW)	Brightness f_L (Typical)	Case
P2000	4.0	50	100	450	171
P2001	3.0	40	120	750	234-02
P2003	4.0	50	100	50	171
P2004	4.0	20	120	750	234-02
P2005	4.0	60	100	—	279

Courtesy Motorola Inc.

Fig. 5-5. Typical light-emitting diodes.

Scan Circuits and Indicators

Table 5-1. Voltage Data for Scan Circuit of Fig. 5-4

TRANSISTORS

Transistor	Emitter (Source)	Base (Gate)	Collector (Drain)	
Q301	0.2	0.2	6.0	MANUAL
	3.0P	3.0P	6.0	SCAN
Q302	5.2	5.9	6.0	MANUAL
	3.0P	3.0P	6.0	SCAN
Q303	0	0.7	0.1	MANUAL
	0	0.1	1.6	SCAN

INTEGRATED CIRCUITS

IC No.	1	2	3	4	5	6	7	8	9	10	11	12	13	14	
IC301	2.0P	2.0P	9.0P	9.0P	2.0P	2.0P	0	2.0P	2.0P	9.0P	9.0P	2.0P	2.0P	3.0P	SCAN
	0.2	3.6	11.0	0.7	3.6	3.6	0	3.6	0.2	11.0	11.0	0.2	0.2	5.2	MANUAL
IC302	2.0P	2.0P	9.0P	9.0P	2.0P	2.0P	0	2.0P	2.0P	9.0P	9.0P	2.0P	2.0P	3.0P	SCAN
	0.2	3.6	11.0	11.0	3.6	3.6	0	3.6	0.2	11.0	11.0	0.2	0.2	0	MANUAL
IC303	3.7P	2.0P	2.0P	2.0P	2.7P	2.0P	0	3.0P	4.0P	3.0P	3.0P	3.0P	4.0P	4.8	SCAN
	1.5	3.6	0.2	0.2	3.6	3.6	0	0.2	1.5	5.9	5.9	0.2	5.9	4.8	MANUAL
IC304	1.6	1.3P	1.8P	1.3P	1.3P	1.3P	0	2.0P	2.7P	2.0P	2.0P	2.0P	2.0	4.8	SCAN
	.1	.2	3.8	3.8	1.5	1.5	0	0.2	1.5	3.6	3.6	0.2	3.6	4.8	MANUAL

NOTE: All voltages are nominal and are measured with a vtvm. SCAN indicates the unit is scanning. MANUAL indicates the unit is not scanning and is stopped at Channel 1. A "P" beside a voltage indicates that the meter reading is pulsating (fluctuating) because the scanner section of the unit is operating.

mination, although the LED may be rated for a maximum current of 60 mA. Application of excessive voltage and/or current will result in permanent dimming or burnout of the LED. When you are troubleshooting this problem, it is advisable to substitute a 39-ohm resistor for the LED and to measure the voltage drop across the resistor. After the cause of excessive voltage has been corrected, the test resistor can be removed and the LED connected back into the circuit.

An LED seldom fails suddenly, unless excessive voltage is applied because of a component defect. Instead, the light output from an LED decreases gradually until it becomes objectionably dim. This is the end of its useful life, and the device should be replaced. It is very poor practice to "doctor" the circuit to increase the applied voltage to the LED. The normal life expectancy of an LED is quite long—as much as 1000 times the life expectancy of an ordinary incandescent lamp. Fig. 5-6 shows how a simple seven-segment readout is connected using LEDs. Some indicators use more than two LEDs connected in series to form each segment. The circuit that is shown in Fig. 5-6 is called the common-anode configuration. Fig. 5-7 shows an exploded view of an LED circuit-board assembly.

Subminiature incandescent lamps, such as the GE "grain-of-wheat" type, are used for channel indication in various scanner-monitor receivers. This kind of lamp has a small voltage drop, such as 1.5 or 2 volts. A typical bulb draws 30 mA at 1.5 volts. Fig. 5-8 shows a scanner printed-circuit board layout (foil side) for a receiver that uses

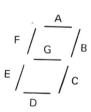

(A) Standard segment arrangement.

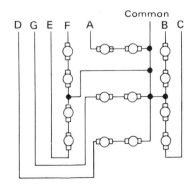

(B) Layout of two LEDs per segment.

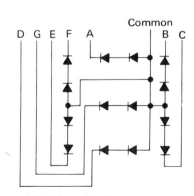

(C) Schematic diagram.

Courtesy Motorola Inc.

Fig. 5-6. A seven-segment LED indicator.

Scanner-Monitor Servicing Guide

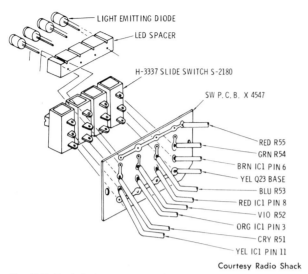

Fig. 5-7. Exploded view of an LED circuit-board assembly.

subminiature incandescent lamps for channel indication. This is an example of an 8-channel scanner-monitor readout. The subminiature lamps have pigtail leads and are soldered into the circuit. Note that although it might appear that a lamp is burned out, the trouble could be due to a defective driver transistor or other component fault. Therefore, to avoid the possibility of replacing a good lamp, the dc voltages for the scanner printed-circuit board should be checked first and compared with the specified values in the scanner-monitor servicing data.

In addition to the arrangement that was shown in Fig. 5-6, other types of seven-segment display devices may be encountered. Display tubes such as the one shown in Fig. 5-9 may have tungsten filaments operating at low voltage. Seven-segment filament-type tubes, such as the RCA Numitron, have a life expectancy of 100,000 hours. A Numitron has seven filament segments mounted on a nonconductive plate and arranged as shown in Fig. 5-10. If the number 8 is to be displayed, all of the segments are energized. The number 4 is displayed, for example, when segments 6, 7, 2, and 3 are glowing. Again, the number 6 employs segments 1, 6, 5, 4, 3, and 7. Other numerals are formed as required by energizing their associated segments. An integrated circuit called a decoder/driver is used to switch the various filaments on and off as required by various displays. This network has four pulse-coded inputs and seven outputs. Fig. 5-11 shows how the decoder/driver is connected to the Numitron tube and shows the pin assignments for obtaining various numerical displays.

Next, it is helpful to observe the scanner printed-circuit board configuration shown in Fig. 5-12. This arrangement uses a seven-segment display indicator. Integrated circuit IC106 operates as a decoder/driver. Transistors Q104 and Q105 operate in conjunction with the decoder/driver to provide a choice of display modes. In other words, if the base of Q105 is connected to the squelch gate by switch S2, as shown in Fig 5-12, the display indicator will glow and display a channel number only when an incoming signal is present. On the other hand, if the base of Q105 is open-circuited by moving switch S2 to the manual (lower) position, the display indicator will then show the number of each channel as it is being scanned during times when there are no incoming signals. As seen in Fig. 5-13, pin 16 on IC106 is the V_{cc} terminal. Thus, C106 can be enabled or disabled in accordance with the output voltage from emitter follower Q104. If the base of Q105 is disconnected from the squelch source, IC106 cannot become disabled.

When switch S2 is thrown to its automatic position, the scan oscillator is enabled. Note that the squelch-gate voltage is applied to the base of transistor Q111. In turn, the collector voltage of Q111 depends upon the squelch-gate voltage level. Thus, Q111 functions as a control switch to start or stop the scan oscillator when the squelch-gate voltage rises or falls. Note that the enable action of Q111 is delayed by the circuitry of scan-delay transistor Q110. In other words, C137 in the delay circuit provides a time constant of four seconds before the enable voltage is applied to the emitter of Q108. Therefore, when an incoming signal stops, a delay of approximately four seconds occurs before Q108-Q109 start oscillating. Then, the scanning system starts to search for other incoming signals, at the rate of approximately 10 channels per second. Note that ferrite beads FB109, FB111, and FB112 are included to act as rf chokes by increasing the lead inductances. In this way, scan-system operation is stabilized.

PRIORITY FUNCTION

A priority oscillator, Q106-Q107, is included in the arrangement shown in Fig. 5-12. When an incoming signal has the monitor locked onto any channels other than priority channel "0," a pulsing noise will be heard in the audio output at four-second intervals. This pulsing noise is caused by operation of the priority oscillator. In other words, the priority oscillator is periodically sampling the "0" channel for the presence of an in-

Scan Circuits and Indicators

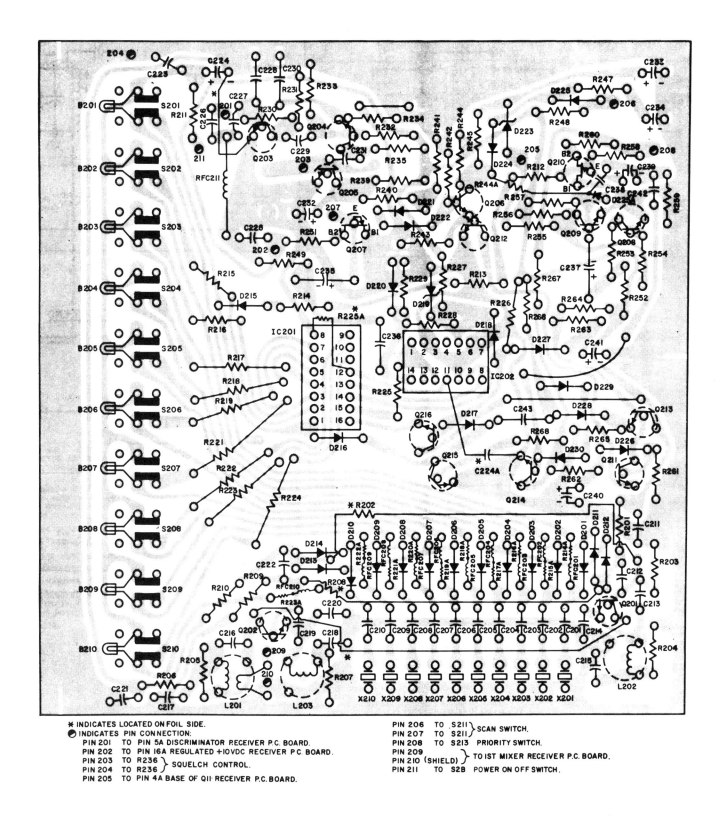

Fig. 5-8. Typical scanner printed-circuit board layout with indicator lamps.

Courtesy Sonar Radio Corp.

Scanner-Monitor Servicing Guide

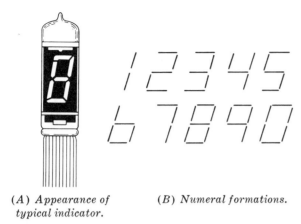

(A) Appearance of typical indicator.
(B) Numeral formations.

Fig. 5-9. Seven-segment display device.

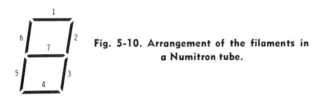

Fig. 5-10. Arrangement of the filaments in a Numitron tube.

coming signal. With reference to Fig. 5-14, when IC101 has stopped counting and is locked onto a channel, the priority oscillator Q106-Q107 will be periodically pulsing an input of each gate in IC102 and IC103 to ground. This causes channel "0" to be sampled because the outputs of three-input NAND gates are driven low, as explained in greater detail subsequently. Integrated circuit IC104 then inverts the low outputs and changes them into high outputs which are coupled in turn to the inputs of IC105 and IC106. Note that IC105 and IC106 are decoder/drivers. When all of the decoder/driver inputs are driven high, a "0" is displayed by the indicator, and the "0" channel crystal is switched into operation.

As noted previously, this sampling occurs every four seconds and has a duration of 20 ms. In case an incoming signal appears on channel "0," the scan circuits will lock onto the signal via the squelch voltage to trigger pin 4 of IC103, which is the input of a 3-input NAND gate. In turn, a voltage is coupled from the "0" output through an inverter (IC104) to produce a "high" level at pin 5 of IC103. This ensures that the monitor will stay locked onto the priority channel until the incoming signal stops. In case of malfunctioning in the scanning system, dc voltage measurements are basic. A high-impedance voltmeter should be used, such as a vtvm or a tvm. Always check the scanner-monitor servicing data and observe the control settings that are specified for voltage measurements.

VOLTAGE SPECIFICATIONS

Fig. 5-15 shows the circuit board layout with specified dc voltages for the scan circuit shown in Fig. 5-12. Two conditions of voltage measurement are included in the data. The voltages shown in ellipses correspond to the mode of operation in which the display indicator shows the number of each channel as it is being scanned. In other words, the base of Q105 in Fig. 5-14 is disconnected from the squelch-gate voltage in this mode of operation. On the other hand, the voltages shown in squares in Fig. 5-15 correspond to the mode of operation in which the display indicator shows a channel number only when an incoming signal is present. For example, the base of Q105 is connected to the squelch-gate voltage in this mode of operation. Note also that the voltages shown in ellipses correspond to the manual mode of operation, and that the voltages shown in squares correspond to the automatic mode of operation.

It is quite feasible to troubleshoot a scan circuit board by measuring dc voltages at the IC pins without knowing what the internal circuitry of

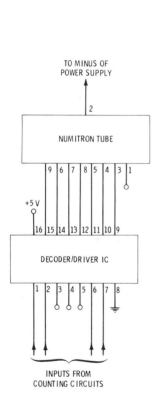

DISPLAY	TUBE PIN ASSIGNMENTS
0	3, 4, 5, 7, 8, 9
1	5, 8
2	3, 4, 6, 7, 8
3	4, 5, 6, 7, 8
4	5, 6, 8, 9
5	4, 5, 6, 7, 9
6	3, 4, 5, 6, 7, 9
7	5, 7, 8
8	3, 4, 5, 6, 7, 8, 9
9	4, 5, 6, 7, 8, 9
PIN NO. 2 COMMON	

(A) Terminal designation.
(B) Pin assignments.

Fig. 5-11. Connections of decoder/driver to Numitron tube.

Scan Circuits and Indicators

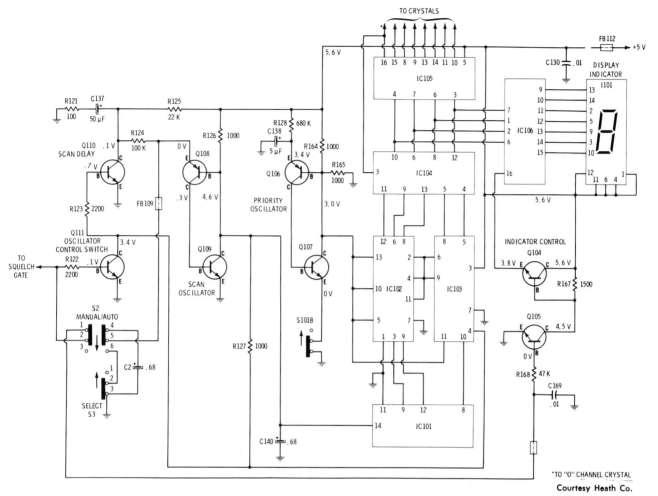

Fig. 5-12. Scanner printed-circuit board configuration, including priority oscillator.

the IC may be. However, these voltages become much more meaningful if the technician recognizes that a particular voltage value corresponds to the state of a NAND gate, an inverter, an input or output terminal of a flip-flop, or whatever the internal circuitry of the IC might be. Thus, an apprentice technician might tend to pay little attention to the type of IC package that is under test, whereas a highly experienced technician generally identifies an IC that is being checked out. In the

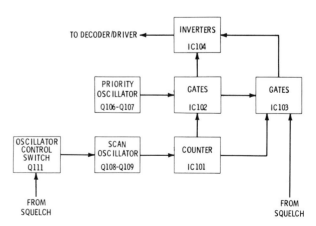

Fig. 5-13. Terminal designations for IC106.

Fig. 5-14. Block diagram for priority oscillator operation.

63

Scanner-Monitor Servicing Guide

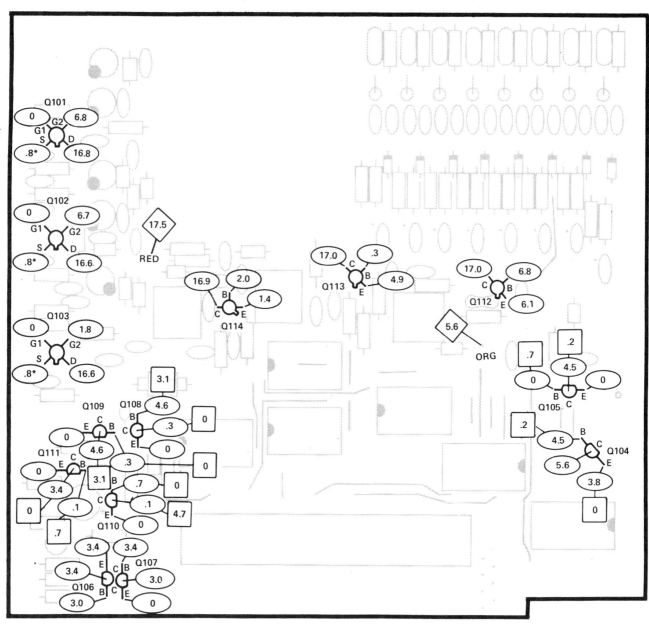

Fig. 5-15. Scan circuit-board layout, with specified dc voltages.

Scan Circuits and Indicators

example under discussion, internal circuitry for IC101, IC102, IC103, IC104, and IC105 is shown in Fig. 5-16. The internal circuitry of IC106 was depicted in Fig. 5-13.

With reference to Fig. 5-16A, IC101 is a counter containing four flip-flops, two NAND gates, and one AND gate. Integrated circuits IC102 and IC103 are packages containing three NAND gates, while IC-104 contains six inverters. Integrated circuit IC-105 is a decoder/driver package; IC106 is also a decoder/driver. Although the details of JK flip-flops, RS flip-flops, and decoder/drivers are not covered in this book, interested readers may refer to *Digital Equipment Servicing Guide*, published by Howard W. Sams & Co., Inc.

TROUBLESHOOTING PROCEDURES

1. Monitor Does Not Scan Automatically

Probable causes for failure of the monitor to scan automatically are as follows:

a. Squelch control turned too far. Readjust setting of squelch control.
b. Marginal supply voltage. Check the voltage; if subnormal, replace battery or troubleshoot power supply.
c. Defective contacts in manual/auto switch. Clean the contacts, or replace the switch.
d. Leaky or open capacitor in scan-oscillator circuit. Measure dc voltages; bridge sus-

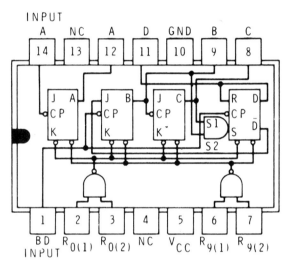

(A) Diagram of IC101.

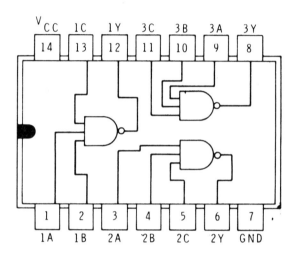

(B) Diagram of IC102 and IC103.

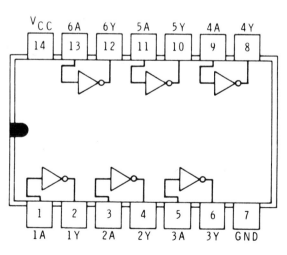

(C) Diagram of IC104.

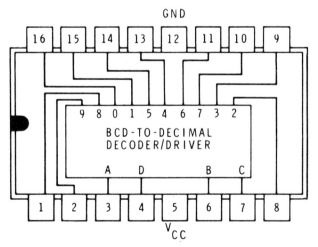

(D) Diagram of IC105.

Fig. 5-16. Internal circuitry of scan-board IC packages.

Scanner-Monitor Servicing Guide

pected open capacitors with a known good capacitor.

e. Defective transistor in scan-oscillator, oscilator-control, or scan-delay circuit. Check dc terminal voltages of suspected transistors.

f. Off-value resistor in oscillator, control, or delay circuit; not likely, but possible.

2. Monitor Scans but Does Not Lock Onto Incoming Signal

Possible causes for the monitor not locking onto incoming signals, although scanning action is normal, are as follows:

a. Crystal lockout switch(es) have been set to lockout position.

b. Defective transistor, such as Q111 in Fig. 5-12. Measure transistor terminal voltages.

c. Malfunction in squelch circuit; check dc voltages in squelch section.

d. Faulty capacitor in squelch line, such as C310 in Fig. 5-4.

e. Off-value resistor in oscillator control circuit (less likely than other component defects).

f. Cold-solder joint or cracked printed-circuit conductor in oscillator control section.

3. Monitor Skips Channel(s) but Has Normal Scanning on Other Channels

Typical causes for the monitor skipping one or more channels but having normal scanning on other channels are as follows:

a. Defective counter IC; check terminal voltages, or make an IC substitution test.

b. Defective gate or inverter IC; check dc voltages, or make a substitution test.

c. Faulty transistor or IC in decoder/driver section; check terminal voltages; make IC substitution test.

d. Integrated circuit making poor contact in socket; check pins for corrosion or mechanical defects.

e. Defective crystal lockout switch(es); make continuity tests in switch circuit(s).

f. Glitches (random pulses) in supply voltage (not likely, but possible); check power supply with an oscilloscope.

4. Scanning Action Normal; Wrong Channel Shown by Display Indicator

Common causes for wrong-channel indication when scanning action is normal are as follows:

a. Display indicator making poor contact in socket; check contacts.

b. Burned-out segment in display indicator; make substitution test.

c. Defective decoder/driver IC. Measure terminal voltages; make substitution test.

d. Faulty display driver transistor (in some configurations). Measure transistor terminal voltages.

e. Malfunction in gate or inverter IC; check dc voltages, or make a substitution test.

f. Cold-solder joint or cracked printed-circuit conductor in display indicator section.

5. Priority Circuit Does Not Function

Probable causes for failure of the priority circuit to function are as follows:

a. Priority oscillator section dead; in this situation, the pulsing sound in the audio output at four-second intervals is missing. Check terminal voltages of oscillator transistors.

b. Defective capacitor in priority oscillator circuit. Check capacitors.

c. Faulty switch contacts in priority oscillator section; check contacts.

d. Off-value resistor in priority oscillator section; not likely, but possible.

e. Crystal in "0" channel does not oscillate when channel is enabled; troubleshoot crystal-oscillator circuit.

f. Poor contact of priority-input IC terminals in socket; check contacts.

6. Delay Circuit Does Not Function

Typical causes for failure of the delay circuit to function are as follows:

a. Open time-constant capacitor, such as C137 in Fig. 5-16. Bridge suspected capacitor with known good capacitor for quick check.

b. Defective delay transistor; measure dc terminal voltages.

c. Off-value resistor in delay circuit (not likely, but possible).

d. Marginal defect in scan-oscillator circuit that prevents scan-delay control. Measure dc voltages in scan-oscillator section.

e. Defective switch in delay circuit; check switch.

7. Does Not Scan but Squelch Operates

Probable causes for failure of the monitor to scan, although the squelch circuit operates normally, are as follows:

a. Defect in auto/manual switch; check switch contacts and action.

Scan Circuits and Indicators

b. Scan oscillator is "dead." Check dc voltages in scan-oscillator section.

c. Faulty IC in scan-oscillator output circuit; measure IC terminal voltages, or make substitution test.

d. Defective decoder/driver IC; measure IC terminal voltages, or make substitution test.

e. Marginal supply voltage to scan board; check voltage and replace battery, or troubleshoot power supply as required.

CHAPTER 6

Audio Section Troubleshooting

Audio amplifiers in scanner-monitor receivers perform the same basic function that they do in standard fm receivers. However, as shown in Fig. 6-1, the audio channel in a scanner monitor includes an audio-squelch gate and is somewhat more elaborate than audio amplifiers in ordinary fm receivers. The audio signal from the fm detector is applied to the base of the audio-squelch gate transistor Q201. Transistor Q201 operates either as a class-A amplifier or as an open switch, depending upon the output voltage level of the emitter of Q206. As noted in Chapter 1, a noise amplifier is included in the i-f section. When there is no incoming signal, the i-f noise level is very high. However, this noise does not normally pass through the audio section because it is automatically blocked by Q201. The circuit action is as follows.

From the noise amplifier, the noise signal is applied to the full-wave rectifier, D201–D202, in Fig. 6-1. In turn, this pulsating dc voltage is amplified by Q204, filtered to practically pure dc by R221–C232, and applied to Q205. In turn, this dc voltage is stepped up and its polarity inverted by Q205. Next, Q206 operates as an emitter-follower current amplifier to drive the audio-squelch gate, Q201. The output level from Q205 also serves to switch the scan oscillator off and on. When an incoming signal is present, the scan oscillator stops and Q201 comes out of cutoff. In turn, Q201 amplifies the audio signal and applies it to the volume control, R1. From R1, the audio signal is coupled to the base of transistor Q301, which is part of a differential amplifier. This differential amplifier also includes transistor Q302 and common-emitter resistor R306 which combines the input and feedback signals.

Next, the signal from the collector of Q301 is applied to the base of the constant-current voltage-amplifier transistor, Q303. In turn, the signal at the collector of Q303 is fed to the base of Q304 and, via diode D301, to the base of Q305. Note that D301 functions as a dc-level shifter, providing a suitable bias voltage for the driver and output transistors. This bias voltage is applied across the Q304 and Q305 base-emitter junctions that are connected in the circuit with D302.

Driver transistors Q304 and Q305 in Fig. 6-1, along with output transistors Q306 and Q307, form a quasi-complimentary output configuration. A positive-going signal applied to the base of Q304 increases its current demand. This in turn increases the current demand of Q306 and increases the current flow through the speaker voice coil. On the other hand, negative-going signals increase the current demand of Q305 and Q307, decreasing the current flow through the speaker voice coil. The driver and output transistors operate in class AB. Since the input and output signals are in phase, any appreciable coupling from the output of the circuit back to the input would cause instability. Therefore, a phase-shifting circuit comprising R316 and C308 is provided to stabilize amplifier operation.

BASIC AUDIO CIRCUIT TESTS

Signal-tracing and signal-substitution tests are generally employed for preliminary localization of a "dead" audio stage or section. Then, dc voltage

Scanner-Monitor Servicing Guide

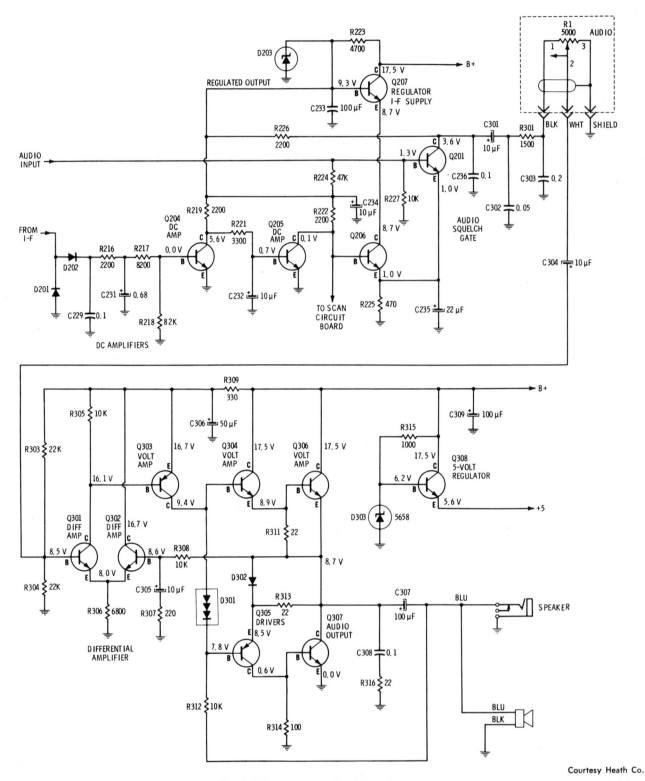

Fig. 6-1. A scanner-monitor audio circuit.

measurements are usually made, supplemented by resistance measurements in case additional data are needed. Sometimes it is helpful to make transistor turn-off and turn-on tests. Since audio systems in scanner-monitor receivers are often dc-coupled, there is much more interaction among stages than in ac-coupled configurations. Therefore, conclusive test data may require more cross-checks than would be necessary in analyzing an equivalent ac-coupled amplifier configuration.

Audio Section Troubleshooting

Since feed-back loops in dc-coupled circuits can be particularly troublesome, these considerations are detailed subsequently.

Signal-injection tests can be made with an audio oscillator. However, a simplified quick check, called a "click test" is generally preferred for preliminary analysis. This quick check consists of producing a transient signal in the audio channel by momentarily short-circuiting the base and emitter terminals of a transistor, as depicted in Fig. 6-2. In this example, transistors Q1 and Q2 normally operate in class A and are forward-biased. Therefore, momentary application of zero bias (removal of the forward bias) will produce a clicking sound from the speaker, provided that the transistor under test and the following stages are working normally. As an illustration, if a clicking sound is heard when a base-to-emitter short circuit is applied to Q303 in Fig. 6-1, it would be concluded that all of the stages following Q303 are operating.

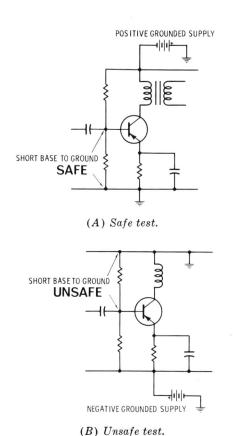

(A) *Safe test.*

(B) *Unsafe test.*

Fig. 6-3. Safe and unsafe base-to-ground click tests.

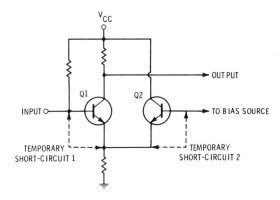

Fig. 6-2. Example of click tests made by short circuiting the base and emitter terminals of a transistor.

As shown in Fig. 6-3, it is sometimes dangerous to short-circuit the base of a transistor to ground for a click test. The danger is due to excessive forward bias that may be applied to the transistor in some configurations, with the result that the transistor would be burned out. However, it is always safe to short-circuit the base terminal of a transistor to the emitter terminal. A similar danger can exist when an audio-generator signal is injected at the base of a transistor, without a series blocking capacitor in the "hot" lead of the generator. For example, with reference to Fig. 6-4, the output resistance of the audio generator is 50 ohms. If a series blocking capacitor were not used, the base of Q5 would be returned to ground through 50 ohms. Effectively, the negative terminal of the battery would apply forward bias to both Q5 and Q6 through 50 ohms, instead of through 2700 ohms. Therefore, both transistors would be very likely to burn out. However, use of a 0.1-μF blocking capacitor ensures that the dc voltage distribution remains undisturbed.

Signal-tracing tests can be made with a conventional signal tracer or with an oscilloscope and a low-capacitance probe. An oscilloscope is more informative than an aural signal tracer because it shows the kind of distortion that might be present, in addition to whether the signal may be stopped at some point. As shown in Fig. 6-5, a common type of distortion is peak clipping. In this example, V_{BE} is subnormal, with the result that positive peaks of the input signal reverse-bias the transistor. In turn, the transistor cuts off, and clipping results. This clipping distortion is passed along into the collector circuit, as shown in Fig. 6-5. An analogous type of distortion occurs when the voltage drop across the load impedance is almost equal to the supply voltage, as depicted in Fig. 6-6. Clipping is due to collector saturation in this case. A common cause of saturation distortion is subnormal $-V_{cc}$, which can result from an increase in the value of R_L, a leaky decoupling capacitor that reduces the source voltage, or other component de-

Scanner-Monitor Servicing Guide

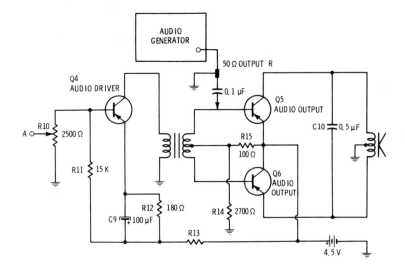

Fig. 6-4. A blocking capacitor is required in the "hot" lead of the generator.

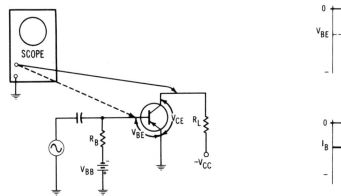
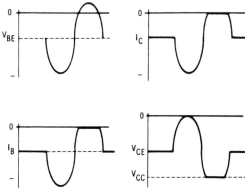

Fig. 6-5. Example of clipping distortion resulting from subnormal base-emitter bias.

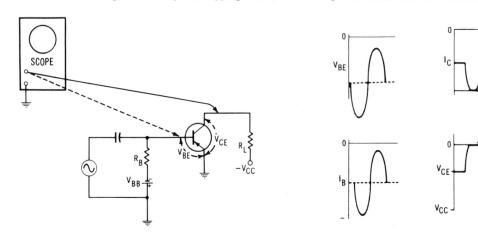

Fig. 6-6. Example of clipping distortion resulting from subnormal V_{CE}.

fect that affects $-V_{cc}$. In summary, distortion will result even when V_{BE} is normal, in case V_{CE} becomes subnormal.

Another basic audio-circuit test concerns identification of a weak stage. This test is not as clear-cut as the identification of a dead stage because the

Audio Section Troubleshooting

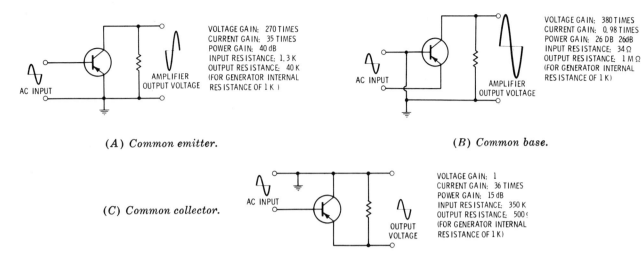

(A) Common emitter.

(B) Common base.

(C) Common collector.

Fig. 6-7. Input/output characteristics for basic amplifier operating modes.

technician must have a general knowledge of what input/output signal ratio to look for. Fig. 6-7 lists representative ratios for the three basic amplifier configurations. When complex audio circuitry is being analyzed, and normal input/output signal ratios are in doubt, it is sometimes possible to make comparative tests against another similar scanner-monitor receiver that is in good operating condition. Note in passing that the input and output resistance values noted in Fig. 6-7 do not refer

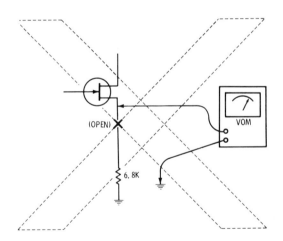

Fig. 6-8. The internal resistance of the vom makes circuit action appear nearly normal.

to ohmmeter measurements. Instead an input resistance value denotes an ac resistance measurement (or impedance). In other words, it denotes the ratio of a small increase in forward-bias voltage to the resulting current increase. An ac resistance value is often very different from a dc resistance value. Interested readers desiring more information on this subject may refer to a text on semiconductor theory.

PRECAUTIONS IN TESTING AUDIO CIRCUITRY

Experienced technicians know that various precautions must be observed when testing solid-state circuitry. For example, a dead stage results from an open source return, as shown in Fig. 6-8. If a vom is used to measure the source voltage, it will appear to be approximately normal. In turn, an apprentice technician might assume that the trouble is not in the stage under test and might overlook the fact that the internal resistance of the vom is substituting for the source resistor. Another type of measurement error is shown in Fig. 6-9. If this kind of base-current measurement is attempted, it would be falsely concluded that the base current is substantially subnormal. In other words, the 2000-ohm internal resistance of the

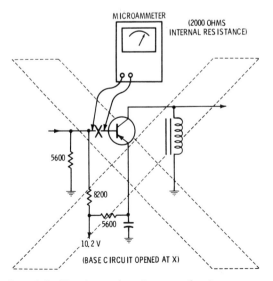

Fig. 6-9. The internal resistance of microammeter seriously disturbs base-circuit action.

microammeter adds excessive resistance in series with the base circuit. A base-current measurement is impractical in this type of circuit.

Ordinary bipolar transistors should not be confused with Darlington transistor configurations. Darlington amplifiers are actually a pair of transistors, contained inside the same case, that are directly connected internally as a single high gain (beta of 500 or more) amplifier. Although a Darlington transistor cannot be tested conclusively with an ohmmeter, it can be checked with a transistor tester for ac beta value. If the device is in normal condition, its beta reading will be in excess of 500. Fig. 6-10 shows the internal configuration of a Darlington transistor. Note that the base-emitter voltage drop is across two series-connected base-emitter junctions. Thus, the normal dc voltage between the base and emitter terminals is double the value for a conventional transistor. Almost all Darlington transistors are silicon types. Accordingly, a base-emitter forward bias of 1.2 volts is typical.

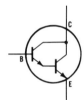

Fig. 6-10. Internal configuration of a Darlington transistor.

DC VOLTAGE MEASUREMENTS

Specified dc voltages for the amplifier in Fig. 6-1 are shown in Fig. 6-11. Before proceeding to pinpoint a defective component by means of dc voltage measurements, it is good practice to check the source voltage. In other words, if the source voltage happens to be subnormal, the dc voltage distribution throughout the amplifier circuitry will be affected accordingly. In some cases, the service data will specify that certain voltages are to be measured with no signal input and that other voltages are to be measured with normal signal input. The signal level will affect various dc voltage values in nonlinear stages, such as class-B and class-AB amplifiers. A class-B stage is practically cut off under no-signal conditions, whereas the stage conducts at maximum under peak signal conditions. In turn, this change in current demand affects the dc voltage drops across resistive circuit components.

Under conditions of instability, the dc voltage distribution in an audio amplifier often changes considerably. As an illustration, if C308 in Fig. 6-1 becomes open, parasitic oscillation will occur. In turn, the associated transistors will be driven into nonlinear regions of operation, and partial rectification will occur with marked shifts in dc voltage values in the circuit. Therefore, it is essential for the troubleshooter to understand the circuitry under test so that the more involved symptoms can be correctly analyzed. Under conditions of intermittent operation, it is often helpful to monitor dc voltages at several points in the audio system. This procedure avoids the possibility that the intermittent may "clear up" as soon as the voltmeter probe is applied to a test point. Intermittent troubleshooting is the most difficult task ordinarily confronted by the technician. If the trouble cannot be pinpointed within a reasonable length of time, it may sometimes be more economical to replace the entire audio-circuit board.

Statistically, audio output stages fail more often than driver stages. This is because the output stage is worked harder than the driver stage, and

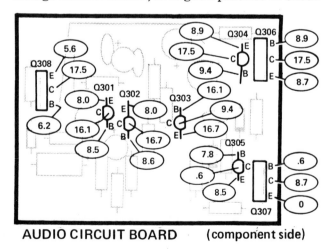

Fig. 6-11. Dc voltage specifications for amplifier circuit shown in Fig. 6-1.

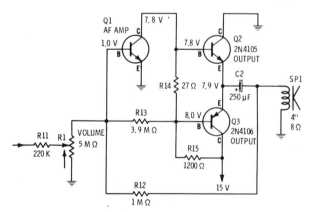

Fig. 6-12. A OTL complementary symmetry audio output circuit.

Audio Section Troubleshooting

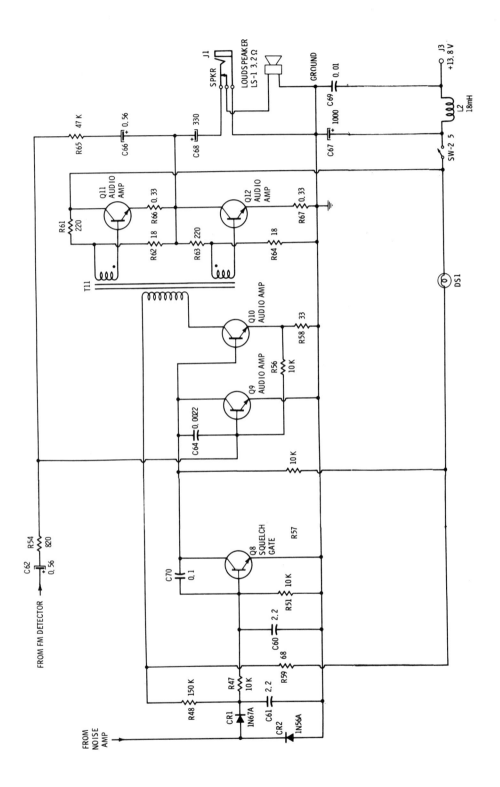

Courtesy E. F. Johnson Co.

Fig. 6-13. Audio-amplifier arrangement with negative feedback from output to input.

handles more power. In the case of an output-transformerless (OTL) stage, a key test is the dc voltage value at the midpoint between the two output transistors, where the speaker is connected. Any dc voltage error here indicates trouble in the output stage most likely one or both transistors are defective. For the circuit shown in Fig. 6-12, the midpoint voltage is specified at 7.9 volts. In case capacitor C2 becomes leaky, the midpoint voltage will be subnormal, although both output transistors are in good condition. The bias voltage in this example is 0.1 volt, indicating that the output stage operates in class B. One transistor is an npn type, and the other is a pnp type; this is called a complementary symmetry configuration. When the circuit is driven to maximum rated output, the forward bias on the output transistors increases substantially.

Next, consider the audio-amplifier arrangement shown in Fig. 6-13. A negative-feedback loop is employed from output transistors Q11-Q12 back to audio amplifier Q9. This is an ac feedback loop. However, if capacitor C56 happens to become leaky, a dc feedback loop is established, and the collector voltage of C12 is bled into the base circuit of Q9. In turn, the bias on Q9 and Q10 shifts. Note that audio output transistors Q11 and Q12 should be a matched pair in order to maintain both bias voltages at the correct value. The output transistor connected directly to the source voltage (Q11 in this example) is the more likely of the two transistors to fail. In case Q11 becomes faulty and Q12 is still all right, it is advisable to replace both Q11 and Q12 with a matched pair of transistors. This procedure ensures that the performance of the output stage will not be below par due to mismatch of the transistors.

TROUBLESHOOTING PROCEDURES

1. No Output

Probable causes for no sound output from the audio section are as follows:

a. Defective transistors, such as Q2 and Q3 in Fig. 6-12. Make a signal-tracing or a signal-injection test. If the minimum output from the audio generator is excessive, use an auxiliary attenuator such as the one in Fig. 6-14.
b. Defect in squelch-gate circuit. Measure dc voltages under signal and no-signal conditions. Check front-to-back ratios of the noise-rectifier diodes.
c. Open speaker voice coil or open coupling capacitor such as C2 in Fig. 6-12.

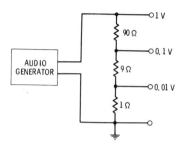

Fig. 6-14. Auxiliary attenuator for audio oscillator.

d. Defective speaker jack, such as J1 in Fig. 6-13.
e. Short-circuited frequency-compensating capacitor, such as C10 in Fig. 6-4.
f. Defective volume control, such as R1 in Fig. 6-1.
g. Short-circuited de-emphasis capacitor, such as C303 in Fig. 6-1.

2. Weak Output

Typical causes for weak sound output from the audio section are as follows:

a. Collector leakage in audio output transistor, such as Q2 or Q3 in Fig. 6-12. Measure terminal voltages of suspected transistors.
b. Subnormal supply voltage.
c. Defective volume control, such as R1 in Fig. 6-12.
d. Leaky de-emphasis capacitor, such as C302 in Fig. 6-1.
e. Open emitter bypass capacitor, such as C57 in Fig. 6-15.
f. Defective driver or output transformer (less likely, but possible).
g. Output transistors may be mismatched; check dc voltages in the output section.

3. Distorted Output

Possible causes of distorted output from the audio section are as follows:

a. Leakage in coupling capacitor, such as C49 in Fig. 6-15, causing incorrect base-emitter bias on next stage.
b. Excessive collector leakage in a transistor, causing shift in normal base-emitter bias. Check transistor terminal voltages.
c. Marginal bias on squelch-gate transistor. Measure bias voltage.
d. Defective decoupling capacitor, such as C67 in Fig. 6-13.
e. Short-circuited emitter-bypass capacitor, such as C9 in Fig. 6-4.

Audio Section Troubleshooting

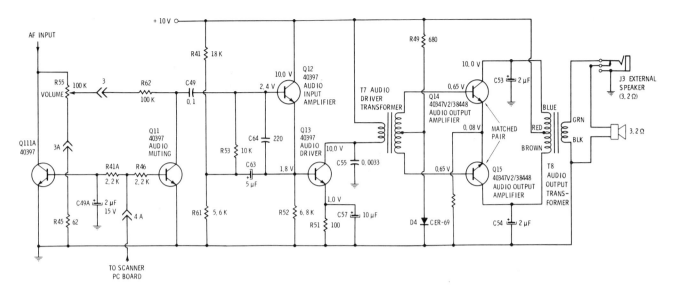

Fig. 6-15. Another audio-section configuration.

Courtesy Sonar Radio Corp.

f. Faulty dc-level-shifter diode, such as D301 in Fig. 6-1.

g. Crossover distortion in output stage due to subnormal forward bias on transistors. (See Fig. 6-16.) Bias resistor(s) may have been overheated due to excessive current demand of a defective output transistor. Note that crossover distortion increases as the audio volume level is reduced.

4. Noisy Output

Common causes for noisy output from the audio section are as follows:

a. Noisy transistor. Temporarily short-circuit the base-emitter terminals of each suspected transistor to determine whether noise will stop.

b. Noise from defective power supply. Check dc-supply-voltage line with scope.

c. Defective volume control. Try short-circuiting the volume control temporarily to determine whether noise will stop.

d. Cold-soldered connection in audio section. Tap various terminals lightly to determine whether noise surges are produced.

e. Defective output transformer (less likely, but possible).

f. Noisy level-shifting diode; check at each end of diode with scope for possible presence of noise spikes.

g. Noisy resistor in preamp or driver stage (not a common fault, but possible).

5. Sudden Shifts in Output Level

Probable causes for sudden shifts in the audio output level are as follows:

a. Defective volume control. Connect an ohmmeter between the wiper arm and one side of the volume control to check smoothness of resistance variation.

b. Cold-soldered connection. Lightly tap various terminals in the audio section to determine whether the symptom is aggravated.

c. Defective transistor. Monitor terminal voltages of suspected transistors.

d. Fluctuation in power-supply voltage. Monitor dc supply voltage with voltmeter.

e. Variations in the audio input signal due to trouble in preceding sections. Monitor audio input level with scope.

f. Marginal squelch-gate action; monitor squelch-gate voltages.

6. Gradual Weakening of Sound Output

Typical causes for gradual weakening of sound output are as follows:

a. Output transistors are drawing excessive current and overheating. Check base-emitter bias on the output transistors.

b. Input signal level is falling due to trouble in preceding sections. Monitor audio input signal level with scope.

c. Power-supply voltage is decreasing. Monitor supply voltage with a voltmeter.

77

Scanner-Monitor Servicing Guide

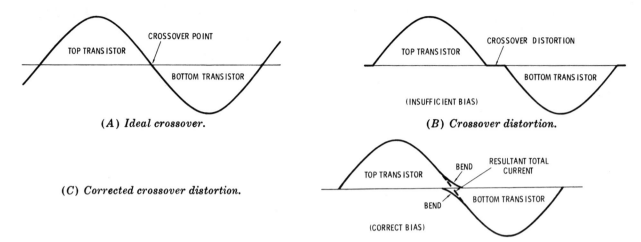

(A) Ideal crossover.

(B) Crossover distortion.

(C) Corrected crossover distortion.

Fig. 6-16. Waveforms showing example of crossover distortion.

   d. Failing diode, such as D4 in Fig. 6-15.
   e. Defective coupling capacitor, such as C2 in Fig. 6-12.
   f. Defective bypass capacitor, such as C9 in Fig. 6-4.

7. Howling or Motorboating

Probable causes of howling or motorboating (putt-putt) in the sound output are as follows:

   a. Defective stabilizing capacitor, such as C308 in Fig. 6-1.
   b. Faulty feedback capacitor, such as C338 in Fig. 6-17.
   c. Open decoupling capacitor, such as C325 in Fig. 6-17.
   d. Open output filter capacitor in power supply.
   e. Malfunctioning integrated circuit, such as IC302 in Fig. 6-17 (less likely, but possible). Substitution test is advisable.
   f. Acoustic feedback from speaker to microphonic transistor, such as Q9 in Fig. 6-13. Tap transistor lightly to determine whether it is microphonic.

8. Raspy Sound in Audio Output

Typical causes for raspy sound output are as follows:

   a. Speaker may be defective. Make a substitution test.
   b. Severe intermodulation distortion may be

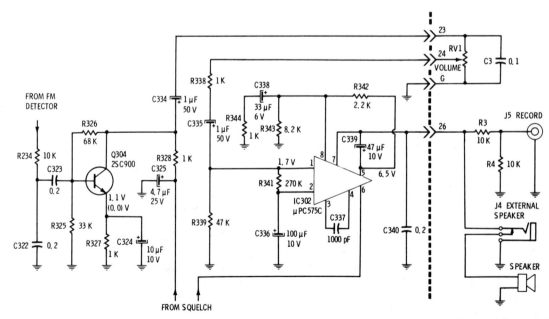

Courtesy Lafayette Radio Electronics

Fig. 6-17. Example of an audio section using an integrated circuit.

present in the audio section. (See Fig. 6-18.) Check base-emitter bias voltages on all of the audio transistors.

c. Serious clipping of the audio waveform could be occurring. (See Fig. 6-19.) Check audio waveform with a scope.

d. Combination of intermodulation and clipping distortion. Incorrect base-emitter bias voltages are likely to be present.

e. If the waveform distortion appears in the input waveform, the trouble will be found in the fm-detector section.

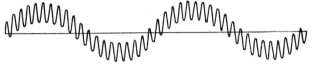

(A) Input waveform.

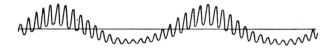

(B) Output waveform.

Fig. 6-18. Example of waveform distorted by intermodulation.

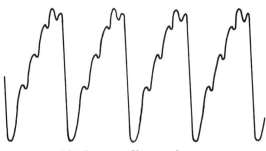

(A) Input audio waveform.

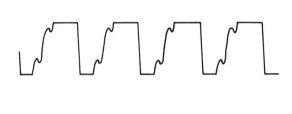

(B) Seriously clipped waveform.

Fig. 6-19. Example of serious peak clipping in audio waveform.

CHAPTER 7

Specialized Scanner-Monitor Operating Features

On occasion, the technician will encounter a scanner-monitor transceiver, as shown in Fig. 7-1. It can be seen from the schematic that the receiver configuration in this example is essentially the same as that discussed in previous chapters. Note that the receiver and transmitter sections use the same antenna, with an antenna relay that is energized by the push-to-talk switch on the microphone. An LC low-pass, pi-filter section is included in the antenna line. This filter suppresses harmonic radiation from the transmitter. Narrow-band frequency modulation (nbfm) is used, with a maximum deviation of 5 kHz. Frequency modulation is accomplished by varying the capacitance across the transmitting crystal that is in use. Thus, crystal Y301 is shown switched into the transmitter oscillator circuit, and frequency modulation results from capacitance variation across Y301 by varactor diodes CR302 and CR303. Note that the audio signal originating at the microphone is stepped up through the speech amplifier (Q205, Q206, and Q207) and is then applied to the junction of CR302 and CR303.

TRANSMITTER ADJUSTMENTS

Before a transmitter is "tuned up," make sure that an antenna with suitable characteristics is connected to the final amplifier (at J1 in Fig. 7-1). In place of the antenna, an equivalent rf load, such as that provided by an rf wattmeter, may be connected to the final amplifier. In the example of Fig. 7-1, a 50-ohm load is required, and this value is standard for practically all scanner-monitor transmitters. Observe that a fail-safe network comprising T301, CR301, Q304, and Q305 is provided for the transmitter. A fail-safe network ensures that the output transistor (Q301) will not be damaged if the rf load is accidentally lost. It operates on the basis of the high swr (standing-wave ratio) that arises from load loss. Fig. 7-2 shows the basic principles that are involved. When a high peak swr occurs, this rf voltage will be rectified by CR301 and applied as cutoff bias to transistor Q305, thereby blocking passage of the rf drive voltage. With no rf drive voltage applied to its base, Q301 cuts off and therefore does not dissipate excessive rf power. (Not all transmitters will contain fail-safe networks.)

A scanner-monitor transmitter should be adjusted with respect to three crystal frequencies. As an illustration, Y301 in Fig. 7-1 might have a frequency near the low end of the band; Y304 might have a frequency near the center of the band; Y306 might have a frequency near the high end of the band. Note that these crystals could be changed as desired after the transmitter has been adjusted. The essential consideration is that the transmitter be aligned with respect to the approximate center frequency of the band. Then, the bandwidth (deviation) is checked with respect to the high end of the band and the low end of the band. If necessary, a slight compromise adjustment is made to ensure that the maximum permissible deviation is not exceeded. The test equipment required for proper adjustment of a scanner-monitor transmitter includes an rf wattmeter, a frequency counter, an fm modulation meter (devia-

Scanner-Monitor Servicing Guide

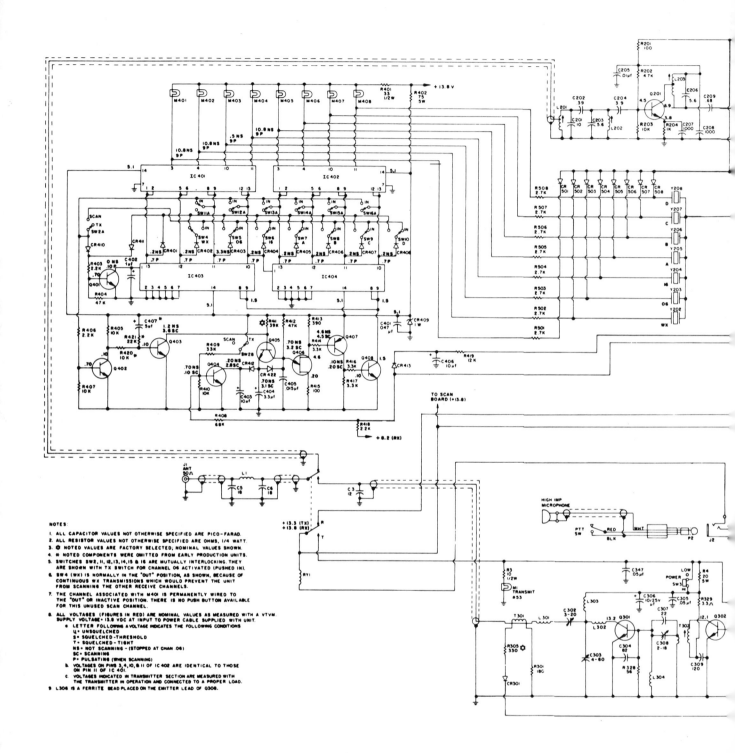

Fig. 7-1. Schematic for a scanner-monitor

Specialized Scanner-Monitor Operating Features

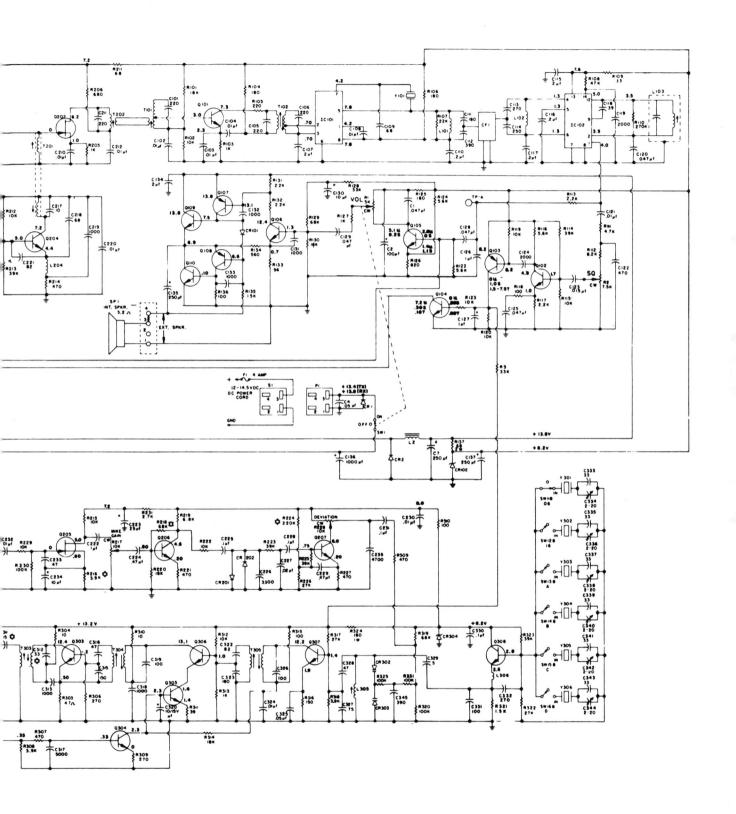

transceiver configuration.

Courtesy Regency Electronics, Inc.

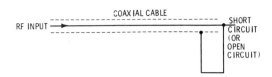

(A) *Rf power flowing from the input end is completely reflected by the short-circuit termination.*

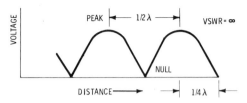

(B) *Power reflected from the load produces nulls along the line causing vswr to be infinite.*

(C) *Load resistor R is unequal to the characteristic impedance of the line.*

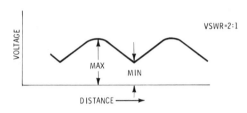

(D) *The vswr is 2:1.*

(E) *Correct load value absorbs all the incoming rf power.*

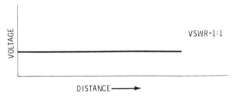

(F) *The vswr is 1:1.*

Fig. 7-2. Basic examples of voltage standing-wave ratios.

tion meter), an audio generator, a tvm, and an oscilloscope.

To begin the transmitter adjustments shown in Fig. 7-1, the rf output tuning capacitors, C302 and C303, are set to their maximum capacitance values. Then, the *netting* capacitors, C334, C336, C338, C340, C342, and C344, are set to midrange. These netting capacitors will eventually be adjusted to bring each crystal precisely on frequency. With the midband crystal switched into the oscillator circuit, and the push-to-talk switch depressed, the dc voltage across R321 normally measures between 2.5 and 3.0 volts. If this voltage is incorrect, a component defect should be suspected in the oscillator section. If the oscillator emitter voltage is correct, the tvm probe is moved ahead to the emitter of Q307, and L305 is adjusted for a maximum reading. In this example, the voltage reading at the emitter of Q307 normally falls between 1.8 and 2.0 volts. Note that failure of L305 to peak indicates that there is a component defect in the associated circuitry.

Next, the tvm is connected at the emitter of Q306 in Fig. 7-1, and the slugs of T305 are adjusted for a maximum reading. The normal voltage reading for this step is between 1.5 and 2.0 volts. Because of circuit interactions, the forego-

ing steps are repeated until no further increase in emitter voltages can be obtained. Note in passing that adjustment of L305 and T305 will shift the oscillator crystal frequency slightly. Therefore, the netting adjustments are deferred until the transmitter tuned circuits have been aligned. Next, the tvm probe is moved to the emitter of Q303, and the slugs of T304 are adjusted for a maximum reading. The normal voltage reading for this adjustment is between 0.4 and 0.6 volt. In case the emitter voltage is incorrect, the associated circuit is checked for component defects.

With reference to Fig. 7-1, tuned transformer T303 is adjusted next, with an rf wattmeter used as an indicator. First, the primary (bottom) slug in the transformer is adjusted roughly midway between the top of the coil collar and the bottom winding. Then, with the mike switch depressed, the secondary (top) slug is adjusted for maximum power indication on the rf wattmeter. A touch-up adjustment of T303 is made subsequently. Sometimes it may appear that the final amplifier (power-output stage) is defective because it is seriously detuned. In such a case, connect the tvm across C309 and adjust the slugs of transformer T303 for minimum collector voltage. Note that when the transmitter is supplying rated rf power output, the voltage across R329 will read between 1.0 and 1.3 volts.

With T303 in approximate adjustment, the tuned circuits for the final amplifier can be brought into preliminary alignment. First, the slug of T302 is set to the center of the coil winding. Then, capacitor C302 is set to near-maximum capacitance. With the rf wattmeter used as an indicator, the following adjustments are made for maximum power output. Capacitor C303 is adjusted for peak reading, then C308 is adjusted for peak reading, and finally C302 is adjusted for peak reading. Because of circuit interaction, these preliminary alignment procedures are repeated for possible improvement in peak reading. Then, T303 is readjusted for maximum rf output. This completes the preliminary alignment of the final amplifier circuit with respect to the midband frequency.

At this point, the bandwidth of the final amplifier may be normal or it may be subnormal. Moreover, the bandpass characteristic might be tilted toward the high end of the band or toward the low end. Therefore, a bandwidth check is made next. This is done by noting the comparative rf output with the low-frequency crystal switched into operation, and then noting the rf power with the high-frequency crystal switched in. Any difference between the two output levels is corrected by adjustment of T303. If the output level at midband is appreciably greater than the output level at either end of the band, it indicates that the bandwidth of the final circuitry is subnormal. This can often be corrected by touch-up adjustment of T302. Finally, capacitor C308 is adjusted for maximum uniformity of rf output at the three spot frequencies.

After the final amplifier is properly adjusted, the crystal *netting* procedure is performed. A frequency counter is used to make precise measurements of the crystal frequencies. The transmitter is worked into an rf wattmeter, as before, or it may be worked into a suitable dummy load. (The transmitter should not be worked into an antenna for purposes of frequency adjustments.) An rf pickup loop of 3 or 4 turns is used as a pickup coil for the frequency counter. The loop is placed near L301 in the final section. Then, the microphone switch is depressed and the first crystal frequency is measured. In turn, the associated netting capacitor (C334, C336, C338, C340, C342, or C344) is adjusted to bring the crystal precisely on frequency. This procedure is repeated for each crystal.

Specialized Scanner-Monitor Operating Features

This netting procedure requires a frequency counter with a range up to 170 MHz. In case a 50-MHz frequency counter must be used, a slightly different procedure is followed. In this situation, the rf pickup is placed near the top of T305, where the prevailing signal frequency is one-fourth of the frequency of the final-amplifier tank circuit. Therefore, the frequency indicated by the counter is multiplied by four. There is a possibility of slight "pulling" action in this second procedure because of the lack of complete buffering between T305 and the crystal oscillator. Therefore, the coupling to the rf pickup loop should be made as loose as practical. The maximum input sensitivity of the counter should be used so that the coupling can be minimized.

Deviation and microphone-gain adjustments are made after the netting procedure is completed. With reference to Fig. 7-1, the microphone-gain control is R217, and the deviation control is R228. As before, the transmitter is worked into an rf wattmeter or into a dummy load. An oscilloscope is connected to the junction of C225 and CR201. Alternatively, the scope can be applied to the cathode end of CR202. The press-to-talk switch is depressed while the technician talks into the microphone at a normal conversational level. Observing the scope waveform, the technician adjusts microphone-gain control R217 to clip approximately 10% of the peaks in the voice waveform. An audio oscillator is then connected at the microphone input. An audio voltage level between 0.5 and 1.0 volt rms with a frequency of 1 kHz is used for the deviation test. The fm deviation meter is coupled to the transmitter tank circuit, and deviation control R228 is adjusted so that the maximum deviation is no greater than 5 kHz. Finally, with the audio input level set to 0.25 volt rms, the deviation should not be greater than 5 kHz.

SELECTIVE-CALL OPERATION

Occasionally, the technician may encounter a transceiver with selective-call facilities. As noted previously, audio tones are utilized to make a scanner-monitor receiver ignore all carriers on a given channel except one that has properly coded tones. Thus, the receiver remains squelched until it is specifically called by the particular transmitter producing the proper coded carrier. Reception of a correctly coded carrier typically unsquelches the audio channel; it may also energize a pilot light or a buzzer to call attention to the transmission. Three principal methods are employed for selective calling with audio tones. One method utilizes an audio tone burst with a frequency in the range from 300 Hz to 3000 Hz to unsquelch the receiver. The burst duration may be from 2 to 5 seconds.

Selective-calling encoders and decoders have the general arrangement shown in Fig. 7-3. In one possible use of selective calling, the operator at the transmitting unit presses a button corresponding to the receiver unit that is desired. In turn, only the selected unit of the four receivers will respond to the transmitted call. Examples of tone frequencies in general use are 700, 725, 2612, and 2704 Hz. However, as noted previously, two-tone sequential (TTS) call signals are usually preferred. The advantage of TTS is that a large number of codes can be formed from a relatively small number of audio frequencies. A TTS encoder generates two audio tones in rapid succession, with an interval of a few milliseconds between the tones. In turn, the decoder at the receiver will respond, provided that the tone frequencies are correct and have the right sequence. The tones must be precisely on frequency and must also have the proper duration.

Audio tones may be selected or generated by means of tuned reeds, which are essentially tuning forks. With reference to Fig. 7-4, the tuned reeds are labelled PEF-1 and PEF-2. The tuned reed is enclosed in a housing approximately one inch by two inches. Observe that the reeds operate as audio filters in the circuit shown in Fig. 7-4. In other words, PEF-1 will pass one particular audio tone to Q303, and PEF-2 will pass another particular audio tone to Q303. A set of four tuned reeds (filters) is provided in this example. The filters are installed as shown in Fig. 7-5 and may be sequenced for the desired mode of selective calling. These modes are listed in the inset on Fig. 7-4. In two-duty operation, a filter-reversing switch is provided, so that the operator can quickly change over from a first-second response sequence to second-first response.

Servicing of a selective-call receiver requires a sequential-tone generator to provide the particular audio tone sequence utilized in the the receiver. A tone-decoder test for the configuration in Fig. 7-4 is made as follows. An ac voltmeter is connected across the speaker terminals, and the volume control is adjusted for a noise-voltage reading of 1 volt. An rf signal is applied to the antenna-input connector from a signal generator at the frequency to which the receiver is tuned. In turn, the output from the sequential tone generator is applied to the external-modulation input jack on the signal generator. The tone output level from the sequential tone generator is adjusted to an amplitude of 0.2 volt p-p, as indicated by a scope. In this example, the time divisions on the sequential tone generator are set for 0.5 second on the first tone, 0.5 second on the interval, and 0.5 second on the second tone. With the signal generator set for fm output, the sequential tone generator is turned on and the deviation of the fm signal is adjusted to 3 kHz for both tones. In normal operation, these tones will be heard from the speaker in the receiver. The monitor switch is set to the ALERT position, and the waveforms of the decoder board are checked with a scope. Correct waveforms for the receiver in Fig. 7-4 are shown in Fig. 7-6.

A block diagram for the selective-calling decoder board is shown in Fig. 7-7. Transistors Q301 and Q302 operate as sequencing gates (electronic switches). In the condition depicted in Fig. 7-4, observe that Q301 is cut off, and Q302 is forward-biased. In turn, tone amplifiers Q303 and Q304 can be energized only by an output from PEF-2 at this time. Transistor Q305 operates as a detector and rectifies the amplified tone output from Q304. In turn, Q306 and Q307 operate as the first timing

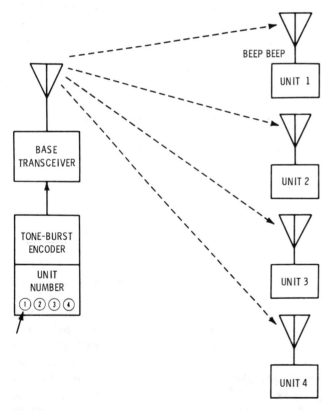

Fig. 7-3. By use of an encoder at each transmitter and a decoder in each receiver, one unit can be called without the message being heard by the other receivers.

gate. This gate triggers one-shot multivibrator IC-301 to open sequencing gate Q301. At this time a second tone may or may not be applied to tone amplifiers Q303 and Q304 from PEF-1. If there is no second tone, the logic section does not change state, and duty-indicator light M1 remains dark. On the other hand, if there is a second tone output from the tone amplifier, the second-tone timing gate pulses the logic sequencer. In turn, the latch enables the astable multivibrator, and duty-indicator light M1 flashes. At the same time, squelch-gate transistor Q104 is brought out of cutoff, and the audio channel is opened.

FAILURE MODES OF DIGITAL ICS

In order to troubleshoot digital logic systems effectively, it is necessary to understand the types of failures that occur. One basic group of malfunctions in these circuits are caused by internal IC failures; another basic group of malfunctions are caused by failures external to the IC. There are four types of defects that can occur internally to an IC. These are:

1. An open bond on either an input or an output.
2. A short between an input or output and V_{cc} or ground.
3. A short between two pins, neither of which is V_{cc} or ground.
4. A failure in the internal circuitry (also called the steering circuitry) of the IC.

In addition to the four internal failures noted, there are four failures that can occur in the circuitry external to the IC. These failures are:

1. A short between a node and V_{cc} or ground (a node is essentially a point at which a circuit branches).
2. A short between two nodes, neither of which is V_{cc} or ground.
3. An open signal path.
4. Failure of a discrete external component.

It is instructive to consider the effect of each of these failures on circuit operation. The first failure noted was an open bond on either an input or an output of an IC. This kind of failure has a different effect, depending upon whether it is an open output bond or an open input bond. With reference to Fig. 7-8, the inputs driven by the open output are left to "float." In transistor-transistor logic (TTL) and diode-transistor logic (DTL) circuits, a floating input rises to approximately 1.4 to 1.5 volts and usually has the same effect on circuit operation as a "high" logic level. Thus, an open output bond in an IC will cause all inputs driven by that output to float to a bad level. As shown in Fig. 7-9, 1.5 volts is less than the high threshold level of 2.0 volts, and greater than the low threshold level of 0.4 volt. The 1.5-volt level is usually interpreted as a high level by the inputs.

In the case of an open input bond inside the IC, as exemplified in Fig. 7-10, the open circuit blocks the input signal from entering the IC. The input on the IC is thus allowed to float and will respond as though it were a static high signal. It is important to recognize that, since the open occurs on the input inside the IC, the digital signal driving this input will be unaffected by the open and will be detectable at the input pin (such as at point A in Fig. 7-10). The effect is to block this input signal inside the IC, and the resulting IC operation will be as though the input were a static high. The second type of defect, a short circuit between an input and V_{cc} or ground, has the effect of holding all signal lines connected to that input or output either high or low—high in the case of a short to V_{cc}, or low if shorted to ground. This situation is shown in Fig. 7-11. In many cases, this condition will cause normal signal activity at points beyond the short circuit to disappear. Thus, this type of failure is catastrophic in terms of circuit operation.

A short-circuit between two pins is not as straightforward to analyze as a short to V_{cc} or ground. When two pins are shorted, the outputs from IC1 and IC2 that are driving those pins oppose each other when one output attempts to pull the pins high while the other output attempts to pull them low, as shown in Fig. 7-12. In this situation, the output that is attempting to go high will supply current through the upper saturated transistor of its "totem-pole" (series-connected transistors) output stage, while the output attempting to go low will sink this current to ground through the lower transistor of its "totem-pole" output stage. In turn, the short will be pulled to a low state by the saturated transistor connected to ground. Whenever both outputs attempt to go high simultaneously or to go low simultaneously, the shorted pins will respond properly. On the other hand, whenever one output attempts to go low, the short circuit will be constrained to the low state.

Next, consider a failure of the internal circuitry (steering circuitry) of an IC, as shown in Fig. 7-13. This defect has the effect of permanently turning on either the upper transistor of the

Scanner-Monitor Servicing Guide

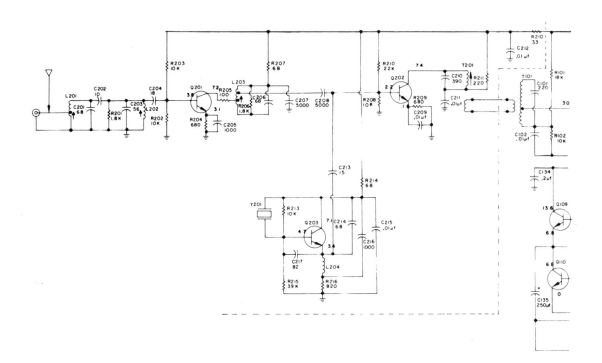

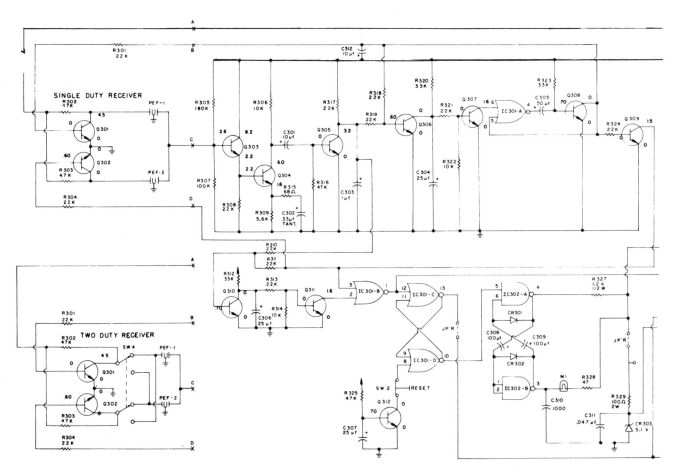

Fig. 7-4. Configuration of a

Specialized Scanner-Monitor Operating Features

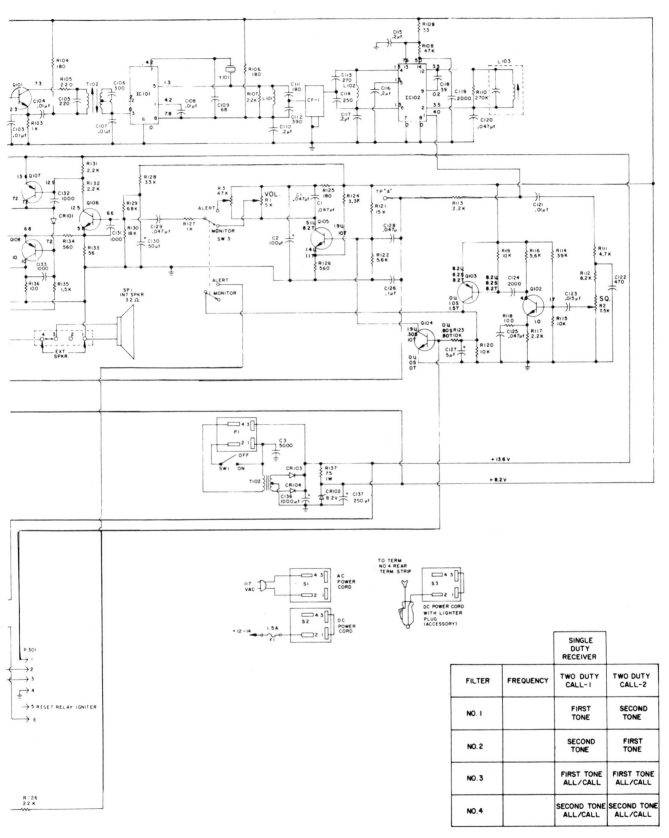

selective-call receiver.

Courtesy Regency Electronics, Inc.

FILTER	FREQUENCY	SINGLE DUTY RECEIVER TWO DUTY CALL-1	TWO DUTY CALL-2
NO. 1		FIRST TONE	SECOND TONE
NO. 2		SECOND TONE	FIRST TONE
NO. 3		FIRST TONE ALL/CALL	FIRST TONE ALL/CALL
NO. 4		SECOND TONE ALL/CALL	SECOND TONE ALL/CALL

Scanner-Monitor Servicing Guide

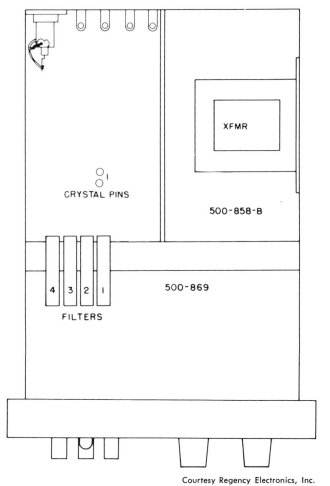

Fig. 7-5. Location of selective-call filters on receiver chassis.

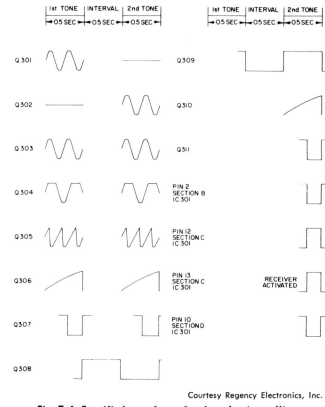

Fig. 7-6. Specified waveforms for the selective-calling decoder board.

"totem-pole" output stage, thus locking the output in the high state, or the lower transistor of the output stage, thus locking the output in the low

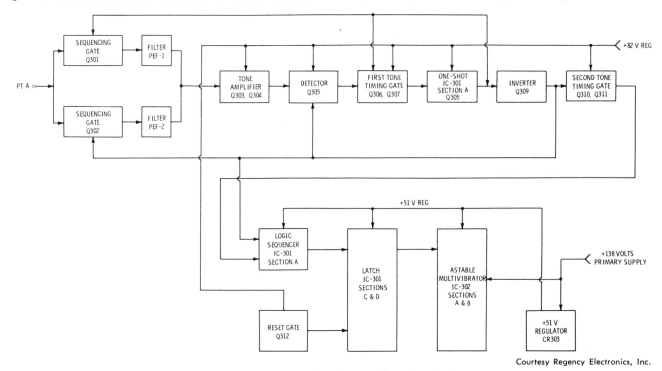

Fig. 7-7. Block diagram for the selective-calling decoder board.

Specialized Scanner-Monitor Operating Features

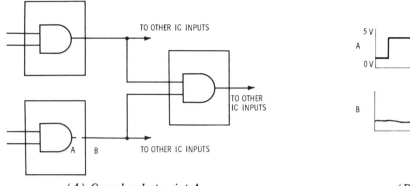
(A) Open bond at point A.

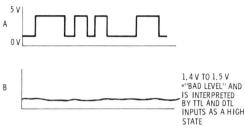
(B) Signals at points A and B.

Courtesy Hewlett-Packard

Fig. 7-8. An example of an open output bond in an IC.

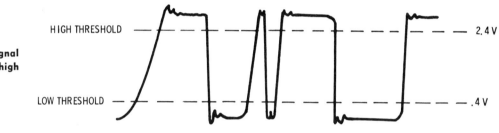

Fig. 7-9. Example of a TTL signal voltage showing the low and high threshold levels.

Courtesy Hewlett-Packard

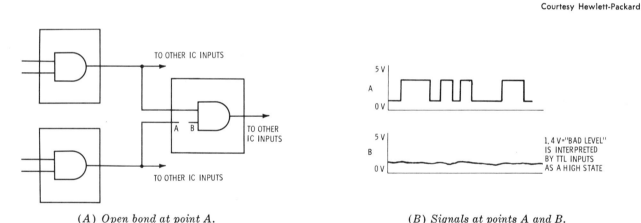

(A) Open bond at point A.

(B) Signals at points A and B.

Courtesy Hewlett-Packard

Fig. 7-10. An open input bond in an IC.

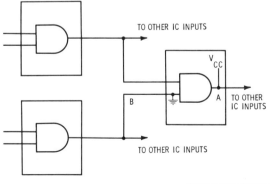

Courtesy Hewlett-Packard

Fig. 7-11. Example of internal shorts from an input to ground and from an output to V_{cc}.

state. This is often called a "stuck-at" condition. It is a failure that blocks the signal flow and has a catastrophic effect on circuit operation.

A short between a circuit connection and V_{cc} or ground external to the IC is indistinguishable from a short internal to the IC. Both conditions will cause the portion of the circuit connected to the node to be either high (for shorts to V_{cc}) or always low (for shorts to ground). When this type of failure is encountered, only a very close inspection of the circuit will serve to determine whether the failure is internal or external to the IC.

91

Scanner-Monitor Servicing Guide

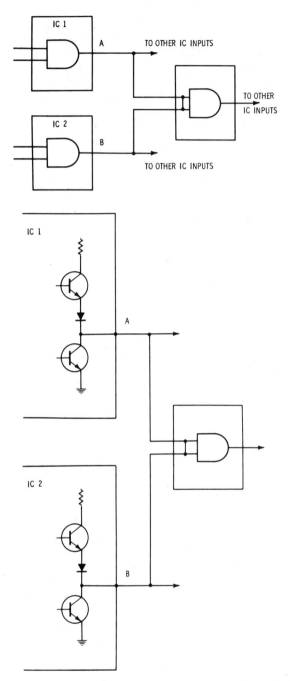

Courtesy Hewlett-Packard

Fig. 7-12. An IC with a short circuit between pins.

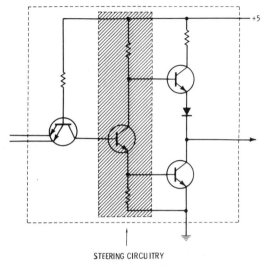

Courtesy Hewlett-Packard

Fig. 7-13. A failure of the internal circuitry of an IC will cause the output to either be static high or static low.

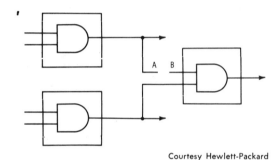

Courtesy Hewlett-Packard

Fig. 7-14. Example of an open in the circuit external to the IC.

TROUBLESHOOTING PROCEDURES

1. Final-Amplifier Transistor in Transceiver Fails

Probable causes of failure of the final-amplifier transistor in a transceiver are as follows:

    a. Fail-safe diode (such as CR301 in Fig. 7-1) has a poor front-to-back ratio.

    b. Final-amplifier transistor (such as Q301 in Fig. 7-1) has excessive collector-base leakage.

    c. Power-supply voltage is excessive; check voltage regulator in automotive electrical system for mobile installations.

    d. Final-amplifier transistor has faulty thermal conduction to heat sink. (See Fig. 7-15.)

2. Drive Failure to Final Amplifier

Possible causes of drive failure to the final amplifier may be localized to any stage between the crystal oscillator and the final-amplifier stage. The dc voltages should be measured first to localize the trouble area (measurement of transistor emitter

As shown in Fig. 7-14, an open signal path in the circuit external to the IC has an effect similar to that of an open output bond inside the IC. All inputs to the right of the open will be allowed to float to a bad level, and they will accordingly appear as a static high level in operation of the circuit. On the other hand, those inputs to the left of the open will be unaffected by the open-circuit condition and therefore will respond normally.

Specialized Scanner-Monitor Operating Features

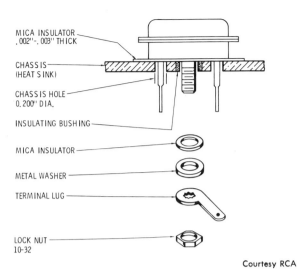

Fig. 7-15. Typical heat-sink assembly.

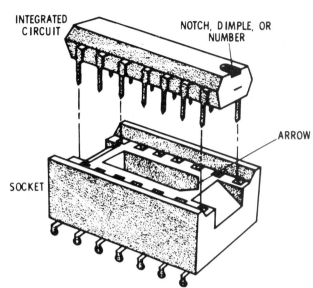

Fig. 7-16. An IC must be correctly oriented and pushed down fully into its socket.

voltages). The following possible component defects should then be checked out:

a. Open capacitor or leaky capacitor in the trouble area.
b. Excessive collector-base leakage in a multiplier or driver transistor.
c. Break in printed-circuit conductor. Flex the circuit board and check the dc voltage distribution.
d. Defective coupling coil or transformer.

3. Carrier Is Not Modulated

When there is carrier output but the carrier is unmodulated, the trouble is most likely to be localized to the speech amplifier. The audio signal from the microphone can be traced with a scope to find out where it stops. Other typical causes of failure to modulate the carrier are as follows:

a. Shorted capacitor in modulator section, such as C345 in Fig. 7-1.
b. Open coupling capacitor between oscillator and modulator, such as C329 in Fig. 7-1.
c. Open coupling capacitor between modulator and speech amplifier, such as C231 in Fig. 7-1.
d. Shorted voltage-divider capacitor, such as C327 in Fig. 7-1.

4. Selective-Calling Decoder Section Is Inoperative

Common causes of failure in the selective-calling decoder section are as follows:

a. Filter sequencing is incorrect; make sure filters are correctly installed.
b. Filter frequencies are incorrect; verify filter frequencies and insert correct filters.
c. Transistor in decoder section may be defective; check the dc voltage distribution and compare it with the values specified in the service data.
d. An IC in the decoder logic section may be defective; check input-output waveforms with scope, or make a substitution test with the suspected IC. (See Fig. 7-6.)

5. Input Terminal of an IC Is "Stuck At" a Bad Level

Identification of the fault when an input terminal of an IC is "stuck at" a bad level is based on the following facts:

a. An open output bond in an IC allows all inputs driven by that output to float to a bad level. This bad level is usually seen as a logic-high state by the inputs, and those inputs driven by the open output bond react as if a static logic-high signal were being applied.
b. An open bond on the input inside an IC has the effect of blocking the input signal from reaching the internal circuitry of the IC and allowing the input to float to a bad level. Even though the input signal is changed from high to low, the IC responds to the bad level as though it were a static high level.
c. Sometimes a replacement IC is accidentally placed in the wrong socket. Check the ICs to make sure they are in the correct sockets.
d. An IC must be correctly oriented and fully inserted in its socket. (See Fig. 7-16.)

Index

A

Afc, 37
Alignment
  rf, 38-40
  sweep, 38-39
AND gate, 9
Antenna
  quarter-wavelength, 13
  scanner-monitor, 13
Audio
  circuits, 69-73
    basic tests, 69-73
    dc voltage measurements, 76-77
  distortion, 29, 71-72
  missing, 14-15, 28-29, 75-76
  output fades, 53
  section, 21, 25-26
  troubleshooting, 69-78
  squelch gate, 69
Automatic
  frequency control, 37
  tuning, 7, 33-34

B

Bandpass filter, i-f, 20
Bandwidth check, 85
Basic troubleshooting, 13-16
Battery, short life, 15-16
Binary-coded decimal decoder, 25
Block diagram, scanner monitor, 8, 9, 17

C

Ceramic filter, 43, 45, 47
Channel
  indicators, 58-60
  lockout switch, 8
  priority, 7
Click test, 71
Clipping, 71-72, 79
Clock, 9
Complementary symmetry configuration, 76
Control, squelch, 21
Crystal(s), 12-13
  calibrator, 38
  certificate, 12-13
  filter, 20, 43
  oscillator, 25, 34

D

Dead receiver, 28
Dc voltage measurements, 76-77
Decoder, 86
Decoder/driver, 25, 60
Delay circuit inoperative, 66
Deviation adjustment, 85
Digital ICs, failure modes, 87, 90-92
Diode switching, 34-35
Display devices, 58-60, 62
  seven-segment, 60, 62
Distorted output, 15, 53, 76-77
Distortion, 71-72
  intermodulation, 78-79
Divide-by-eight counter, 23
Double conversion, 8, 35
DTL, 87
Dual-trace oscilloscope, 28

E

Electronic tuning, 32-34
Encoder, 86

F

Fading audio output, 53
Filter
  ceramic, 43, 45, 47
  crystal, 22, 43
  LC bandpass, 47
Final amplifier
  drive failure, 92-93
  transistor failure, 92
Fixed-tuned rf circuit, 32-34
Flip-flops, 23
Frequency
  counter, 38
  response curves, 38-39, 43

G

Gate, 9
  AND, 9
  squelch, 21
General Arrangement of a scanner monitor, 8-10
Grain-of-wheat lamp, 59

H

Harmonic
  third, 36-37
  ninth, 36
Hi-lo multimeter, 49, 51
Howling, 78

I

I-f
  bandpass filter, 20
  circuitry, 43-53
  frequencies, 43
  response curve, 43
  stages, 20-21
Indicates wrong channel, 66
Indicators, 58-60
Integrated circuit, 45
Interference, 15
Intermediate-frequency network, 20-21
Intermittent operation, 40, 53
Intermodulation distortion, 78-79
Inverter, 25

J

JK flip-flop, 23

L

Latch, 23
LC bandpass filter, 47
LEDs, 58-59
Light-emitting diodes, 58-59
Lockout switch, 8, 55
Low sensitivity, 29, 40, 52-53

M

Manual tuning, 7
Mixer, 35-36
Monitor does not scan, 29
Motorboating, 78

N

Narrow-band fm, 13
National Weather Service, 8
Nbfm, 13
Negative feedback, 25, 75

Index

Netting
    capacitor, 84
    procedure, 85
Ninth harmonic, 36
Noise, 53
    amplifier, 69
    in mobile operation, 16
Noisy reception, 41, 77
Numitron, 60, 62

O

Oscillator
    circuitry, 34-37
    crystal, 34
Oscilloscope, 28

P

Pi filter, 81
Plug-in antenna, 13
Pocket scanner, 7-8
Power-supply section, 26
Priority
    channel, 7
    function, 60, 62
    oscillator, 25

Q

Quarter-wavelength antenna, 12-13
Quasi-complementary output, 26
Quieting-sensitivity test, 44-45

R

Raspy sound, 78-79
Response curve, 38-39, 43
Rf
    alignment, 38-40
    amplifier, 33
    circuits
        electronic tuned, 32-34
        fixed tuned, 32-34
    response curve, 38-39
    sensitivity, 31-32
    stages, 17, 20
    tuning, automatic, 33-34
Roof-mount antenna, 13

S

S curve, 43-44
Scan
    circuits, 55-67
        basic tests, 56-58
        voltage specifications, 62-65
    clock, 55-56
    inoperative, 53, 65-67
Scanner-monitors, 7-13
    antenna, 13
    audio section, 21, 25-26
    basic-troubleshooting, 13-16
    block diagram, 8, 9, 17
    crystal-oscillator section, 25
    general arrangement, 8-10
    i-f stages, 20-21
    physical layout, 10-13
    power-supply section, 26
    rf stages, 17, 20
    specialized features, 81-93
    squelch section, 21
    scanning section, 21-25
    transceiver, 81-83
    troubleshooting procedures, 28-29, 40-41, 52-53, 65-67, 72-78, 92-93
    tuning circuits, 31-41
Scanner, pocket, 7-8
Scanning
    section, 21-25
    waveforms, 24
Selective-call operation, 85-87
Selective-calling decoder inoperative, 93
Semiconductor testing, 47-52
Sequential-tone generator, 86
Seven-segment display device, 60, 62
Signal
    injection tests, 26-27
    tracer, 27
    tracing tests, 27-28
Single channel dead, 40
Skips channels, 66
Slope detector, 43
Sound, raspy, 78-79
Squelch
    circuit, 8-9
    control, 21
    gate, 21
    inoperative, 53
    section, 21
Standing-wave ratio, 81, 84
Steering circuitry, 87
Stuck-at condition, 90-91, 93
Superheterodyne, double-conversion, 8
Suppressor-resistor cable, 16
Sweep alignment, 38-39
Swr, 81, 84

T

Testing, semiconductor, 47-52
Third harmonic, 36
Transmitter adjustments, 81-85
Transistor
    tester, 49-51
    turn-off test, 50-52
    turn-on test, 51-52
Tripler, 36-37
Troubleshooting
    audio section, 69-78
    procedures, 28-29, 40-41, 52-53, 65-67, 75-78, 92-93
Trunk-lid mount antenna, 13
TTL, 87
TTS, 86
Tuned reeds, 86
Tuning
    automatic, 7
    circuits, 31-41
    manual, 7
TVRS cable, 16
Two-tone sequential, 86

U

Uhf band, 31
Unstable operation, 29

V

Varactor, 34
Vhf band, 31
Voltage specifications, scan circuits, 62-65
Volume changes, 40-41

W

Waveforms, scanning, 20
Weak
    audio, 72-73, 76, 77-78
    output, 15